# STUDIES IN HISTORY, ECONOMICS AND PUBLIC LAW

Edited by the

**FACULTY OF POLITICAL SCIENCE
OF COLUMBIA UNIVERSITY**

**NUMBER 345**

## OCCUPATIONAL DISEASES

IN RELATION TO COMPENSATION AND
HEALTH INSURANCE

BY

ROSAMOND W. GOLDBERG

# OCCUPATIONAL DISEASES

## In Relation to Compensation and
## Health Insurance

BY

ROSAMOND W. GOLDBERG, Ph.D.

NEW YORK
COLUMBIA UNIVERSITY PRESS
London: P. S. King & Son, Ltd.
1931

To
J. A. G.

# PREFACE

THE problem of sickness in its various phases has been attracting considerable attention of recent years. This may perhaps be due to the rising interest in all aspects of preventive medicine as distinguished from curative medicine and to the mounting costs of sickness. For the wage-earner serious illness approaches the stage of being almost a calamity, as with limited means the problem of obtaining adequate medical care, necessary after-care and industrial rehabilitation, becomes a most serious one. And yet, each year large numbers of industrial workers are compelled to face the consequences of serious and debilitating illness, lack of income during such time of involuntary unemployment, needs of dependents and subsequent readjustments. Many of these workers are not even entitled to the benefits of the workmen's compensation laws, for their ailments are to be classified as occupational diseases which are but infrequently covered by such laws. Similar situations were faced for many years by workers in other countries where industry was highly developed. Measures have been taken by many of the European governments to meet the problem of industrial hazards created by rapid industrialization, through instituting systems of sickness or health insurance. Some thought has been given to the subject in this country. The purpose of this volume is to review the various hazards to which many workers are being regularly exposed, and to determine whether a system of health insurance can be applied to meet the large number of cases of occupational diseases and industrial poisoning.

The author wishes to take this means of expressing her appreciation to Professor Robert E. Chaddock for his many valuable suggestions in the preparation of this volume, as well as to Professor Samuel McCune Lindsay for his assistance and for many other courtesies shown to her during residence at Columbia University.

# TABLE OF CONTENTS

## CHAPTER III

### HAZARDS IN METAL, CHEMICAL AND MISCELLANEOUS INDUSTRIES

## CHAPTER IV

### REGULATION AND PREVENTION OF OCCUPATIONAL DISEASES

## CHAPTER V

### WORKMEN'S COMPENSATION LEGISLATION AND JUDICIAL DECISIONS RELATING TO OCCUPATIONAL DISEASES

## CHAPTER VI

### HEALTH INSURANCE AND OCCUPATIONAL DISEASES

# CHAPTER I

## INTRODUCTION

THE problem of occupational diseases did not originate in modern times. While the nature of the various diseases to which industrial workers are exposed can be determined by a study of modern industrial conditions, yet the earliest evidences of the results of deleterious occupational environment reach far back into history. In the great civilizations of the past whether in the East, West or in Europe generally, there was sufficient concentration of labor to produce the intensest forms of the maladies classed by Pliny as the " diseases of slaves." Some of the most injurious processes known to us are extremely ancient; among these may be included lead and quicksilver mining, the potters' craft, textile processes of preparing and weaving asbestos and flax, etc.[1] Hardly a generation passed that was not exposed to industrial diseases, the extent of the suffering depending upon the stage of industrial development of the people at a given time.

That lead poisoning has existed for a very long time is apparent from the uses to which we know lead has been put in the past. When Rome was at the height of her power, and the supply of water for drinking and bathing purposes became a pressing requirement, the problem was partly solved by the construction of large aqueducts, which in turn made the use of lead pipes for domestic distribution more or less of a necessity. The fairly extensive use of lead for this purpose undoubtedly affected the workers in the lead

[1] Oliver, Thomas, *Dangerous Trades* (London, 1902), p. 25.

industry.   That the people suffered from drinking lead-contaminated water is evident from the condemnation of the metal by a famous Roman architect of the day.[1]   There is also satisfactory evidence that the Romans carried on considerable lead-mining operations.   Bars of pig-lead have been found in Derbyshire, England, stamped with the imperial arms of Rome, indicating that the Romans worked the mines and smelted the ore.[2]

### MIDDLE AGES

Georgius Agricola was one of the early observers of the hazards to which miners were exposed.   In his work *De Re Metallica* he wrote as follows:

It remains for me to speak of the ailments and accidents of miners, and of the methods by which they can guard against these, for we should always devote more care to maintaining our health, that we may freely perform our bodily functions, than to making profits.   Of the illnesses, some effect the joints, others attack the lungs, some the eyes, and finally some are fatal to men.

Where water in shafts is abundant and very cold, it frequently injures the limbs, for cold is harmful to the sinews. To meet this, miners should make themselves sufficiently high boots of rawhide, which protect their legs from cold water; the man who does not follow this advice suffers much ill-health, especially when he reaches old age.   On the other hand, some mines are so dry that they are entirely devoid of water, and this dryness causes the workman even greater harm, for the dust which is stirred and beaten up by digging penetrates into the windpipe and lungs, and produces difficulty in breathing, and the disease which the Greeks calls asthma.   If the dust has corrosive qualities, it eats away the lungs and implants consump-

[1] Kober, G. M. and Hayhurst, E. R., *Industrial Health* (Philadelphia, 1924), p. 412.

[2] Oliver, *op. cit.*, p. 283.

tion in the body; hence in the mines of the Carpathian Mountains women are found who have married seven husbands, all of whom this terrible consumption has carried off to a premature death. At Altenberg, in Meissen, there is found in the mines black pompholyx which eats wounds and ulcers to the bone; this also corrodes iron, for which reason the keys of their sheds are made of wood. Further, there is a certain kind of cadmia (probably cobalt) which eats away the feet of the workmen when they have become wet, and similarly their hands, and injures their lungs and eyes. Therefore for their digging they should make for themselves not only boots of rawhide, but gloves long enough to reach the elbow, and they should fasten loose veils over their faces; the dust will then neither be drawn through these into their windpipes and lungs, nor will it fly into their eyes. Not dissimilarly, among the Romans the makers of vermillion took precautions against breathing its fatal dust.[1]

Mercury has long been recognized because of its deleterious effects upon those whose occupations brought them into direct contact with the metal. The first case of occupational poisoning from mercury was reported about 1557 by Fernel, who cited the case of a gilder who became deaf and dumb after mercury poisoning. Forestus, in 1602, while investigating occupational mercurialism, reported a gilder who became a paralytic as a result of exposure to the fumes of the metal.[2] Martin Pansa, a pupil of Agricola, pointed out that mercury, pyrites, cobalt and other metals may liberate poisonous vapors.[3]

It is not till the 17th century that really concrete information is available regarding the condition of mercury miners.

[1] Agricola, Georgius, *De Re Metallica*, translated from the first Latin edition of 1556 by Hoover, Herbert C. and Lou H. (London, 1912), p. 214.

[2] International Labour Office, *Occupation and Health*, Brochure no. 3 (Geneva, 1925), p. 2.

[3] International Labour Office, *Occupation and Health*, Brochure no. 94, 1928, p. 1.

Walter Pope, about 1665, reported that on visiting mercury mines he noted among the workers engaged in extracting the metal certain symptoms of poisoning, such as trembling, paralysis and marked malnutrition.[1]

Data reported in regard to the prevalence of occupational diseases at different periods of time must be interpreted with considerable caution. All observers were not well trained medical or scientific men, and with the constantly changing industries and exposure of workers to varying conditions, errors were quite apt to be made. In the case of anthrax, for instance, it has been contended that in ancient Greece the term was applied to smallpox, though its relation to the disease in man following the manipulation of animal skins was not recognized until comparatively recent times.[2]

Such, in brief, is the outline of the early condition of industry, and the status of the working classes. Out of the earlier experiences came a new view with regard to industrial occupations. There arose an organized type of study not only of the industries, but also of the occupational hazards and diseases. Though in early historical periods there had been a recognition of the causal connection between employments and certain ailments, there had never been an organized attempt to collect this valuable information in order that it might be made of general use.

### MODERN PERIOD

It was not until Bernardino Ramazzini undertook to gather the scattered information and to add thereto his own careful observations that there became available a treatise on the subject of occupational diseases.[3] Ramazzini was

[1] International Labour Office, *Occupation and Health*, Brochure no. 3 p. 2.

[2] Oliver, *op. cit.*, p. 622.

[3] Ramazzini, Bernardino, *On the Diseases of Artificers*. Translated by James, R. (London, 1750).

born at Capri, in the Duchy of Modena, in 1633, and became Professor of Physic at the University of Padua. In 1700 he produced his memorable treatise with the title: *De Morbis Artificum Diatriba*. Though, as was the custom, the book was written in Latin, its merits were speedily recognized and it found translators in different countries. In each European state successive writers referred to his treatise, and supplemented its facts by others derived from their own research.[1]

With the production of this valuable work began a new era in the development of the modern outlook on the industrial diseases. Not only was there a recognition of an increasing number of new industrial diseases, but also, thought began to be given to methods of prevention. This new movement of which Ramazzini was the forerunner, has been of sufficient import to designate him as the father of modern industrial medicine. The fact that stands out in his work is the humane and sympathetic attitude towards the laborer. Some of those among whom he reported occupational and industrial diseases were metal diggers, gilders, surgeons, chemists, potters, coopers, tinsmiths, glass-makers, glass-grinders, painters, printers, blacksmiths, stone cutters, and many others.[2]

### DEFINITION AND CLASSIFICATION OF OCCUPATIONAL DISEASES

In attempting to define occupational diseases one is met with a number of difficulties. Many students of the subject have sought to formulate all-inclusive definitions which might meet not only industrial conditions of the present but also of the future. The citation of some of these definitions

[1] Arlidge, J. T., *The Hygiene, Diseases and Mortality of Occupations* (London, 1892), p. 6.

[2] Ramazzini, *op. cit.*, pp. 35-185.

may help to indicate the problems involved in attempting to draw up a standard definition. According to Theodor Weyl [1] the term occupational disease should be applied not only to the diseases gradually brought on by the activities of the individual worker, but to any frequent morbid condition in a determined class of occupation. Akin to this definition is one which states that " occupational diseases are the outcome of long exposure to noxious influences during work and occur either exclusively or with particular frequency among the workers in a specific industry." [2]

According to another group of experts, an occupational disease is one " connected with the following of certain occupations and arising therefrom, whereby it is regarded as a further occupational risk, or one which is a continuing and lasting cause." [3] Hayhurst [4] defined an occupational disease as an affliction which is the result of exposure to an industrial health hazard, with possible exposures to more than one hazard with corresponding complicated afflictions.

A few definitions even refer to the " diathesis ", though it is obvious that there can be no such connection in many forms of occupational disease. Certain of the above definitions are patently open to criticism. It is not necessary to follow an occupation for a long time to contract an occupational disease. On the contrary, such diseases frequently strike down beginners, young apprentices and even temporary workers. In a more general way it is often a matter of great difficulty to differentiate with certainty and precision

[1] Weyl, Theodore, *Handbuch der Arbeiterkrankheiten* (Jena, 1908), p. xxi.

[2] International Labour Office, *Studies and Reports*, Series M. no. 3, 1925, p. 7.

[3] *Ibid.*, p. 7.

[4] Hayhurst, E. R., *Occupational Diseases; Definition, etc.*, U. S. Bur. Lab. Stat. Mo. Lab. Rev., vol. xxix, no. 1, July, 1929, p. 29.

between the parts played by various factors which may have acted upon the organism, and to ascribe the disease partially or exclusively to occupational causes.

It is equally difficult even for highly competent industrial physicians to determine the noxious effect of one of the substances enumerated in a schedule of occupational diseases when the injury or disease is due to two or more substances, both of which are present at the same time at the place of work. The frequency of such cases of " mixed poisoning " is far greater than is generally recognized, particularly in the modern chemical industry.

Occupational diseases and chronic poisoning in particular, proceed by short attacks sometimes followed by fairly long periods of what is to all appearances at least, normal physiological equilibrium. The idea of " slow " poisoning is therefore too limited, for cases occur in which the absorption of the poison in large quantities takes place over a relatively short period of time, while they are not usually classed as accidents in the legal sense.[1] An example of this type of poisoning is that of carbon monoxide, whose onset of symptoms may be as sudden as a stroke of lightning. It has happened in mine accidents that the dead men were found sitting in natural positions, perhaps with their lunch in their hands or with their picks as if they had just paused in their work.[2] So, too, in cases of anthrax infection death may occur in twenty-four hours.[3] There are other industrial poisons having a sudden onset, all of which indicates the inadequacy of the particular definition in question.

Diseases of occupation have also been defined as " in-

[1] International Labour Office, *Occupation and Health*, Brochure no. 3, p. 10.

[2] Hamilton, Alice, *Industrial Poisons in the United States* (New York, 1925), p. 381.

[3] Kober and Hayhurst, *op. cit.*, p. 671.

juries and disturbances of health contracted in industrial pursuits and other vocations of life, as a result of exposure to toxic agents, infectious organisms or other conditions inimical to health." [1]    Those diseases may be acute or chronic, and vary in intensity and duration from the acute and fatal attacks of asphyxia caused by suffocating gases, to the slow, insidious forms of industrial tuberculosis.

Every disease recognized as particularly frequent in an occupation or in a profession ought to be considered as an industrial disease to the extent that it is clearly due to the risks in the trade.    By the term occupational or industrial disease is meant, in brief, the direct effects of a particular trade in which a person is engaged.    In some instances, there are also included maladies that are the result of pathological alterations of structure, indirectly induced by the occupation.    Colic, for example, might be regarded as the direct effect of working in some way with lead, and yet this is unaccompanied by a structural alteration in the wall of the intestine; on the other hand kidney disease in the file-cutter is very slowly developed, and although, as regards its production, there are other causes in operation than lead, still the kidney lesion is believed to be a remote or indirect consequence of plumbism. [2]

A further index of some of the difficulties involved also relates to the matter of lead.    A number of cases have been recorded in medical journals of acute abdominal pain occurring in painters, which had been regarded as lead colic; but when the patients died the post-mortem examinations showed that death was due to a small ulcer of the bowel or to the inflammation of the appendix, which a surgical operation might have cured.    On the other hand, workers exposed to lead have been operated upon for similar symptoms

[1] *Ibid.*, p. 1.
[2] Oliver, *op. cit.*, p. 15.

by competent surgeons for appendicitis when the malady was lead colic.[1]

One of the serious aspects of the entire problem rests in the ever-changing chemical and allied industries, in which new products are constantly coming from manufacturing plants, and at the same time are creating new and frequently baffling industrial hygiene problems. These conditions tax the ability of the most skilled physicians and laboratory technicians, and are raising new problems before many of the older ones have been solved. Thus, most experienced physicians are acquainted with the various forms of dermatitis, but every physician is not aware that such diseases may be caused by certain exotic timbers. Further, physicians are familiar with jaundice, though few are probably acquainted with the fact that it is the principal symptom of tetrachlorethane poisoning.[2]

### SUMMARY

The development of industry in the United States has brought about many industrial and social problems, chief among which have been the hazards to the health and lives of the workers, and the problem of providing for their medical care. For the worker, serious illness of himself or of a member of his family presents many perplexing questions. Proper medical and nursing care, after-care and industrial rehabilitation become costly matters to the man of limited income. Large numbers of industrial and other workers are compelled to face the consequences of serious and debilitating illness, lack of income during the period of disability and subsequent industrial and other adjustments. Many of these wage-earners are not even entitled to the benefits pro-

[1] Oliver, *op. cit.*, p. 805.

[2] International Labour Office, *Occupation and Health*, Brochure no. 3, p. 11.

vided by workmen's compensation laws, for their ailments are often classified as occupational diseases, which are but infrequently covered by such laws.

The numerous hazards and occupational diseases to which workers have been exposed, have been found to a large extent in dusty trades, in metal and chemical industries, and in certain miscellaneous industries. In those industries in which dust is a primary factor in causing disability, the affections of the workers are usually related to the respiratory tract, with tuberculosis standing out as one of the most important consequences of exposure to certain kinds of industrial dusts. In industries in which there is exposure to certain metals, lead stands out as a major cause of disability, not only because of its definite deleterious effect upon the human organism, but also because of its widespread use in industry. Other outstanding metals causing illness of workers include arsenic, brass, iron, mercury and zinc.

The hazards in the chemical industries are usually found in the toxic gases, vapors and fumes. In addition to disability caused by such agents, workers are exposed to various other occupational diseases such as anthrax; those resulting from high temperatures and intense illumination; from work in compressed air; contact with radioactive substances; and a number of other working conditions, many of which only come to be recognized after considerable damage has been done to the health of exposed workers.

In the United States, as in other countries that are highly industrialized, the health problems arising out of exposure to various substances and dangerous working conditions, have been recognized for some years. In order to cope with these problems, four main methods have been pursued, namely, (1) reporting of occupational diseases; (2) regulation of working conditions; (3) prohibition of the use in industry of certain substances; and (4) the enactment of workmen's compensation legislation. In the pages that follow it will be indicated that despite these four methods which have thus far been applied to the control of occupational diseases, such industrial hazards still remain a menace to the health and lives of the workers. European countries have for many years attempted to control the incidence of occupational diseases and to provide for the injured workers, not only by granting compensation through workmen's compensation laws, but also by providing for systems of both voluntary and compulsory sickness insurance for the industrial workers.

In this country, the several states, as well as the federal government, have enacted legislation dealing with various aspects of the problem of regulating working conditions which might result in the development of occupational diseases. Mention will be made of laws regulating the reporting and investigation of occupational diseases; control of employment in mines, factories, mills and mercantile establishments; prohibition of the employment of women and children in certain hazardous industries; regulation of lead industries; control of working conditions in bakeries and food establishments; regulations concerning the use of wiping cloths; work in compressed air; etc. The utilization of industrial codes for the regulation of certain occupations is also considered.

Another method by which attempts have been made to deal with the problem of occupational diseases is through workmen's compensation legislation. It will be noted that by placing upon the employer the partial cost of work accidents, and in some instances, of occupational diseases as well, a method of regulation is exercised which functions to check in part at least the industrial accident and disease rates.

There are two general methods of providing for occupational diseases through compensation legislation: the so-called blanket method and the schedule method by listing specific occupational diseases for which compensation may be granted. Both of these methods will be presented, as well as a representative group of judicial decisions affecting cases in which occupational diseases were involved. These decisions note the complexity of the present system of handling such cases, as well as some of the injustices which follow.

Some of the questions that arise are, whether the development of health insurance legislation in the United States on the model of European systems though adapted to American conditions, would be desirable, and whether it would be advisable to substitute a system of compulsory social insurance for individual and voluntary provision against sickness among the industrial workers.

Under the present methods of compensating for occupational diseases the difficulty in most of the cases is to prove that the disability is due to the nature of the employment and not to some other factor. The frequently slow development of occupational diseases makes the placing of the responsibility upon the individual and last employer a difficult matter.

The purpose of this study is to determine whether a system of health insurance can be applied to meet the problems created by the large number of cases of occupational diseases and industrial poisoning. The material consulted in the preparation of this study has included: standard texts on special phases of the problem; technical journals; official state and federal reports; reports of governmental commissions; reports of the International Labour Office; state and federal laws and state industrial codes; reports and statistics of insurance companies; judicial decisions of the higher courts in various states; and miscellaneous studies and reports issued by authorities on various subjects relating to this study.

# CHAPTER II

## HAZARDS IN DUSTY TRADES

### INTRODUCTION

THE importance of dust as a factor in occupational mortality has attracted the attention of every authority on occupational diseases from Ramazzini to Sir Thomas Oliver. It requires no extended consideration to prove that human health is much influenced by the character of the air breathed and that its purity is a matter of very considerable sanitary and economic importance. Aside from the risk of exposure to so-called air-borne diseases, the pollution of the atmosphere by organic and inorganic dusts is unquestionably the cause of a vast amount of ill-health and premature mortality, chiefly among men and women engaged in the many indispensable trades and occupations that minister to human needs. The sanitary dangers of air contaminated by disease-breeding germs are possibly not so menacing as generally assumed, while the destructive effects of the dust-laden atmosphere of factories and workshops are a decidedly serious menace to health and life.[1]

One of the most authoritative and extended discussions of the entire subject of the inhalation of dust, its pathology and symptomology, with special reference to the dusty trades, is by Dr. J. T. Arlidge, who, in 1892, published his treatise on *The Hygiene, Diseases and Mortality of Occupations.* He noted that:

Few indeed, are the occupations in which dust is not given off,

[1] *U. S. Bureau Labor Statistics Bulletin no. 231*, June, 1918, p. 26.

and in none can it be absolutely harmless, for the lung tissue must be so much the worse, and less efficient for its purpose, in proportion to its embarrassment by dust. Its disabling action is very slow, but it is ever progressive, and until it has already worked its baneful results upon the smaller bronchial tubes and air cells and caused difficulty of breathing, with cough and spitting, it is let pass as a matter of indifference—an inconvenience of the trade.[1]

It might be well to note that a misconception of the ability of the human being to become accustomed to repeated health hazards, is responsible for a large percentage of preventable sickness and mortality which is not always recognized as having a common origin in exposure to occupational hazards. By an occupational or industrial hazard is meant any condition or manner of working that is unnatural to the physiology of the human being so engaged.[2]

One of the major difficulties in dealing with the matter of dust as a hazard in industry lies in the fact that dusty operations are so general and practically unavoidable in all the principal industries, that a precise line of demarcation between injurious and non-injurious processes on account of the dust hazard cannot be drawn, even on the basis of a thoroughly scientific investigation.[3] It is because of this and similar inherent difficulties that various attempts have been made through legislation and regulation to try to control and minimize the hazards to which the workers are patently exposed, and, at the same time, to attempt to provide for the workers who become incapacitated because of the definite hazards.

[1] Arlidge, *op. cit.*, p. 244.

[2] Hayhurst, Emory R., "A Survey of Industrial Health Hazards and Occupational Diseases in Ohio," *Ohio State Board of Health Report,* Feb., 1915, p. 10.

[3] *U. S. Bureau Labor Statistics Bulletin no. 231*, p. 437.

### CLASSIFICATION OF DUSTS

There is not at the present time a general agreement as to the best scheme for the classification of dusts. Perhaps the broadest classification is that suggested by Hoffman [1] who divides the dust groups according to the trade groups in which they fall—metal dust, mineral dust, mineral industries, vegetable fiber dust, animal and mixed fiber dust, organic dust, mixed organic and inorganic dust. Thompson [2] simplifies this classification, making five groups: mineral, metallic, vegetable, animal, and mixtures of two or more of the preceding groups. While either of these classifications is satisfactory from an academic point of view and covers the whole field, they are both unsatisfactory for practical use, because, except in rare instances, a single variety of dust unmixed with some other is never found.[3]

Oliver's [4] original classification was simpler than either of the groupings given above. He divided dusts into the mechanical and irritative, and the chemical and toxic or caustic types. It is possible that this general classification had suggested itself to him because of his belief that the most important factor in the causation of lung diseases due to dusts was the hardness of certain dust particles and the mechanical irritation which results from their presence in the lungs. Collis favors the suggestion that the basis of classification should be the pathological effects caused by exposure to the dust; and Pancoast and Pendergrass [5] follow a similar plan

[1] *Ibid.*, pp. 40-41.

[2] Thompson, W. Gilman, *Occupational Diseases* (New York, 1914), p. 384.

[3] *U. S. Public Health Bulletin no. 176*, "Health of Workers in Dusty Trades," 1928, p. 4.

[4] Oliver, Sir Thomas, *Diseases of Occupation* (New York, 1916), p. 242. *U. S. Public Health Bulletin*, 1915, vol. xxviii, p. 258.

[5] Pancoast, H. K. and Pendergrass, E. P., *Pneumoconiosis (Silicosis)*, (New York, 1926), p. 11.

in their classification of dangerous dusts, dividing them into: (1) those dangerous because of poisonous action, such as lead; (2) those which cause only irritation of the respiratory tract, resulting in such conditions as chronic bronchitis and asthma; and (3) those which tend to produce pulmonary fibrosis and so may predispose to tuberculous affections, and especially pneumonia.

Another classification divides the dusts according to their physical properties. These are: (1) cutting dusts, composed of minute crystalline particles with sharp cutting edges, which include sand, stone, lime, steel, glass, etc.; (2) irritant dusts, which include wood, ivory, textile fabrics, such as wool, fluff, silk, cotton, flax, shoddy and hair; (3) inorganic poisonous dusts derived from certain chemicals, as mercury, copper, arsenic, lead, etc.; (4) soluble saline dusts derived from soluble crystalline substances, including sulphate of iron, sulphate of copper or sodium sulphate; (5) organic poisonous dusts, as tobacco, hellebore and certain organic drugs; and (6) obstructive and irritating dusts as soot, coal, flour and starch.[1]

Whatever plan may be followed as a basis for classification, one is struck, in attempting to apply it, by the wide variation both in the nature and what is more important, in the prevalence of the respiratory diseases which are caused by exposure to the different types of dust.[2] Although the classification by Hoffman does not take sufficient cognizance of mixed dusts which are the ones most frequently found, yet it is seemingly the broadest in its application and apparently best suited for a general discussion of the problems to be considered in the following pages.[3]

[1] *N. Y. State Dept. Labor Special Bulletin no. 90*, Dec., 1918, p. 3.

[2] *U. S. Public Health Bulletin no. 176*, 1928, p. 5.

[3] For a list of about 700 hazardous occupations see: *U. S. Bureau Labor Statistics Bulletin no. 306*, April, 1922. Classifies abnormalities of work-

### THE OCCUPATIONAL MENACE OF METALLIC DUST

The continuous and considerable exposure of workmen to the inhalation of metallic dust in its various forms is generally recognized by medical and other authorities on occupational diseases, as probably the most serious health hazard with particular reference to a material increase in liability to pulmonary tuberculosis and non-tuberculous respiratory diseases. The term "metallic dust" is for practical reasons limited to finely comminuted particles of iron, steel, brass, gold, silver, bronze, lead, arsenic and other metallic substances. Some of these are exceedingly common in connection with industrial processes, while others are rarely met with. Occasionally the pathological aspects of the problem are complicated by chemical considerations, aside from the physical or mechanical properties of the several varieties of metallic dust referred to.

In the vast majority of mechanical operations in which metallic dust is generated there is more or less intermixture with particles of mineral dust, which quantitatively may exceed in importance the ascertainable presence of metallic dust. On account of the heavier weight of metallic particles, the relative degree of air pollution in factories, workshops, mines, etc., where metallic dust is generated is considerably less than the corresponding amount of air impurities resulting from atmospheric pollution by mineral dust. The injurious consequences of industrial dust exposure are, broadly speaking, proportionate to the amount of dust inhaled into the lungs. However, important exceptions to this conclusion are brought out by the consideration in detail of the several kinds of metallic dust, of which, perhaps, lead and

ing conditions as follows: abnormalities of temperature; compressed air; dampness; dust; extreme light; infections; poor illumination; repeated motion, pressure, shock, etc.; the poisons; occupations exposed to specific skin irritants.

arsenic are the most harmful, on account of the additional liability to industrial poisoning.[1]

Hoffman, in his industrial and occupational classification of the kinds of unavoidable dust exposure, with special reference to the mortality from pulmonary tuberculosis, includes the following occupations as those in which there is exposure to inorganic metallic dust: artificial flower makers, brass workers, chippers at blast furnaces and steel rolling mills, compositors and typesetters, cutlery makers, die setters and sinkers, electrotypers and stereotypers, engravers, filers, gold beaters, grinders, card grinders in cotton mills, manufacturing jewelers, polishers, buffers, finishers, pressmen and press feeders, sand blasters, saw filers, solderers and toolmakers.[2] In discussing the occupational menace of metallic dust as indicated in the above listed occupations, it is presumed that it will not be necessary to enter into a detailed description of the hazards in each of the occupations. By selecting certain of the outstanding occupations listed a better picture may be obtained of the important industries in which a large number of workers are engaged, and in which it is possible to determine the extent of the hazards with a degree of accuracy.

Foundrymen and molders are to a considerable degree exposed to the continuous inhalation of both metallic and mineral dusts. While the proportion of metallic dust is quantitatively small, it is probably the more injurious of the two, although its effects are modified by the relatively larger amount of mineral dust. The industry is varied, of large extent, and widely distributed throughout the United States. The conditions affecting health naturally vary, and chiefly so on account of the metal used in casting, which may be iron, steel, brass, etc.[3]

[1] *Ibid., Bulletin no. 231*, pp. 51-52.

[2] *Ibid.*, p. 432.

[3] *Ibid.*, p. 283.

The grinding trade includes a large variety of employments, of which metal grinding, either by dry or wet process, is hygienically as well as industrially the most important. The grinding of metal probably involves as much exposure to decidedly health-injurious conditions as does any other employment, if not more so. Chiefly as the result of the inhalation of relatively large quantities of fine metallic dust, and not inconsiderable quantities of fine mineral dust, the mortality from pulmonary tuberculosis in this occupation is decidedly above the normal for occupied males generally. While the sanitary and other conditions injuriously affecting the health of metal grinders never has been so notoriously bad in the United States as in England or in continental Europe, the observed mortality from this occupation as carried on in this country fully warrants the most serious conclusions as to the health-injurious effects of this group of employments.[1]

Core and mould makers are exposed to the inhalation of large quantities of dust, especially in sifting the sand for the cores and in dusting the completed moulds with powdered charcoal and graphite. It is by no means infrequent to observe red-hot cast-iron stoves giving off deadly coal gas or, what is even worse, open wood or coke fires, provided not so much for the comfort of the workers as for the drying-out of the moulds and repaired fire-clay linings for the " pouring metal pots ".[2]

Metal polishers are engaged in work which to a considerable extent is similar to that in which grinders are engaged. The mortality rate for these workers is similarly high, with

---

[1] *Ibid.*, p. 82; also Kober and Hayhurst, pp. 188-189. For early history see Thackrah, C. T., *The Effects of Arts on Health*, 2nd ed. (London, 1832), p. 94 and Young, Thomas, *Historical Treatise on Consumptive Diseases* (London, 1815), p. 370.

[2] Kober and Hayhurst, *op. cit.*, p. 184.

a special excess of mortality from respiratory diseases.  The chronic inflammatory conditions produced, together with exposure to dampness and other injurious factors, naturally favor infection with the tubercle bacillus.[1]  The wet process in metal polishing, in addition to the danger of spray inhalation, involves other risks.  While dust production is reduced, the constant throwing off of water saturates the clothing, floors and air, and predisposes to colds and rheumatic conditions.[2]

File cutting appears to be one of the most dangerous of all dusty occupations.  Besides the inherent hazard from the fine steel particles, there is the added danger from the particles of lead used in the process of hardening the files by passing them through kettles of molten lead.  The files are usually brushed by the bare hands in order to remove the particles of lead, and thereafter the fingers may be moistened to enable the operative to obtain a better hold upon the file during the process of cutting.[3]  In England it has been found that the file cutters registered a mortality of 85 per cent in excess of the average.[4]

In the cutlery industry the dangers to the health of the workers have been found to be due to mineral and animal dusts, fumes and gases, and to faulty illumination.  In a study of the hazards in several large factories in Massachusetts it was found that in general the dust hazard was greatest in three departments: (1) handle, sawing and milling; (2) grinding; and (3) polishing and glazing (including buffing); that fumes and gases presented a serious hazard in

[1] The Registrar-General's Decennial Supplement, *England and Wales,* 1921, pt. ii, pp. 23-24.

[2] Kober and Hayhurst, *op. cit.,* pp. 188-189.

[3] *Ibid.,* p. 191.

[4] Registrar-General's Decennial Supplement, *England and Wales,* 1921, pt. ii, p. 22.

the forge department where lead, antimony, copper and tin were used. It was further found that the high mortality associated with the cutlery industry was unnecessary, and can be reduced almost to the minimum by suitable attention to dust removal; also, that the process of polishing, grinding and buffing of small articles as in the cutlery industry is a greater hazard from dust to the worker than the same work on larger articles, because of the necessity of close proximity of the worker to the grinding, polishing or buffing surface.[1]

The industrial health hazard to workers in the iron and steel industries has been largely of the respiratory tract. In an analysis of the death rate from pneumonia among iron foundry workers, the Metropolitan Life Insurance Company found that for the three-year period from 1922 to 1924 inclusive, 15.9 per cent of the deaths among iron foundry workers insured by the Company were due to pneumonia, while 7.7 per cent among all occupied males were due to this same disease. The age distribution of the disease showed the same trend, being higher for each age of the chief working periods of life, from 15 to 64 years. The study further disclosed that of 74 occupational classes analyzed 15 had death rates in excess of that for all occupied males. Some of the most significant groups having high pneumonia rates were iron, steel and foundry workers, and polishers on iron and steel products.[2]

The manufacture of jewelry in all its branches involves a large variety of manipulations, including the melting and refining of small quantities of the precious metals, and the handling, shaping, cutting and polishing of precious stones. Most of the articles made by jewelers are of small dimensions, and require painstaking care in handling and continu-

[1] *U. S. Bur. Labor Stat. Monthly Labor Review*, vol. vi, no. 1, Jan., 1918, pp. 181-184.

[2] *Ibid.*, vol. xxvi, no. 6, June, 1928, p. 51.

ous eye-straining attention in shaping and polishing. Aside from the use of gold and silver, many other metals and mineral substances are employed. The work of the jeweler is naturally an indoor occupation, involving a stooping position. The workshops are generally small and the ventilation is usually poor. The dust generated in the processes of hammering, cutting, shaping, grinding, polishing, etc. is considerable but very minute. The health problem is complicated by the almost universal use of the blowpipe apparatus and gas for heating purposes.[1]

Regarding health hazards and mortality, it may be interesting to note that the experience of an American industrial insurance company shows that 812 deaths among jewelry workers from all causes, 39.5 per cent were due to diseases of the lungs and respiratory passages. The conclusion of an investigation based on these figures was that jewelers are subject to a decidedly excessive mortality from pulmonary tuberculosis at ages under 45 and particularly between 15 and 34 years. As regards mortality, the same results have been noted among the jewelry workers in Vienna and Berlin.[2]

Brass casting, founding and moulding are arduous occupations, exposing the worker to the inhalation of considerable quantities of mineral dust more or less mixed with metallic ingredients. Whether brass dust, as such, is more injurious than the dust of iron and steel, for illustration, has not been fully determined. The exposure of the brass workers to dust inhalation is only one of a number of specific factors in a trade decidedly injurious to health and life, and of these mention may be made of the exposure to fumes and vapors generated in the smelting processes. Brass founders' ague is a well-defined occupational disease, the symptoms

[1] *U. S. Bur. Labor Stat. Bulletin 231*, pp. 117-118.

[2] International Labour Office, *Occupation and Health*, Broch. no. 77, 1927, pp. 2-3.

of which are tightness of the chest with indefinite nervous sensation, followed by sweating and fever. Zinc and other fumes inhaled are the chief causes of this ailment, and it is quite probable that the lung injury resulting from the inhalation of fine particles of metallic dust is a material contributory cause in brass founders' ague.[1] Oliver found that the ague in this industry is not a disorder for which brass workers consult a physician, because they realized its transitory nature; but they did come in large numbers to hospitals in order to be treated for bronchial affections.

Ramazzini, in noting the hazards of copper and tin workers, wrote as follows:

The workmen whose business it is to melt and hammer copper and tin are exposed to the same misfortunes that the former are for the subtle atoms exhaled from the copper while it is frequently heated for easier expansion, enter the lungs, raise a dry cough and corrode the texture of the windpipe in the lungs. They likewise produce a discolored complexion. The nature of these particles lodged in the copper is set in a clear light by the beard and hair of the workmen which in the workhouses, become green. As for those who work in tin, they usually are attacked by the same symptoms that the melters and grinders of lead, of which sort are the potters.[2]

In summarizing the occupational hazards inherent in exposure to metallic dust, it is of course realized that there are a number of other industries, as previously listed, in which metallic dust causes diseases of the respiratory tract, as well as other diseases. However, an analysis of many industries would in several instances merely duplicate information already noted in regard to the steel, iron, brass and other trades and occupations. The statistical department of the Metropolitan Life Insurance Company has recently prepared

---

[1] Oliver, *op. cit.*, p. 458.

[2] Ramazzini, *op. cit.*, p. 67.

an analysis of the percentage of deaths due to tuberculosis of the respiratory system from all causes in specified occupations, and covering the experience of white male policyholders of industrial insurance, 15 years and older, between the years 1922 and 1924. In this study it is indicated that workers in certain industries have an excessive mortality, among such workers being miners; pottery workers; cutlers; grinders; brass foundry workers and polishers of iron and steel products.[1]

### MINERAL DUST

Mineral dust exposure is the most common in the stone industry, among potters, in cement manufacture, and in mining. Mineral dust varies widely in its mechanical and chemical properties, and much more so than is the case with the different varieties of metallic dust. The quantitative degree of dust exposure is also decidedly greater in the case of mineral dust, which frequently contaminates the entire atmosphere in a finely comminuted form for prolonged periods of time. According to Hoffman,[2] whose classification of dusts has been previously mentioned, the following are the workers who are exposed to mineral dusts: asbestos workers; brick, tile, terra cotta factory hands; core makers; color mixers; glass blowers; other workers in glass factories; lacquerers, japanners and enamelers; lime, cement and gypsum factory workers; lithographers; marble and stone workers; mica workers; mirror makers; moulders; paint factory workers; paper hangers; plasterers; potters; whitewashers; miners of asphalt, bauxite, coal, copper, gold, silver, graphite, lead, zinc, mica, phosphate, quicksilver, spar, sulphur and others not specified.

[1] *Statistical Bulletin*, Metropolitan Life Ins. Co., vol. ix, no. 6, June, 1928, p. 4.

[2] Hoffman, *U. S. Bur. Labor Stat. Bull. no. 231*, p. 164.

The hazards inherent in dusty trades in which there is exposure to mineral dust were recognized by many early writers, including Ramazzini. Writing upon the diseases of stone cutters, he made the following remarks:

Stone hewers, statuaries and stone cutters when polishing and cutting the rock, oftentimes suck in by inspiration the sharp, tough, small splinters or particles which fly off so that they are usually troubled with a cough and some of them turn asthmatic and consumptive. In dissecting the corpses of such artificers, the lungs have been found stuffed with little stones. Several stone cutters who died from asthma were opened and in their lungs were found such heaps of sand that in running the knife through the pulmonary vesicles it seemed that one was cutting some sandy body.[1]

Beddoes, writing in 1799, also noted that stone workers were especially liable to develop pulmonary consumption as a result of the dust which entered their lungs.[2]

Dust may be inhaled, or ingested or it may affect the skin, the eyes and the ear canals. The daily subjection of an individual to dust for more than brief periods at a time is always damaging. The skin and the eyes may become physiologically inured to mineral dust, but such is not the case in regard to the internal organs. Dusts in general produce a chronic catarrh of the respiratory and digestive organs. All diseases of the lungs that are due to dust are known as pneumoconiosis (lung dust disease), and there are several classifications of this condition, depending upon the causative type of dust such as iron giving rise to siderosis; flint and other stones to silicosis; and coal dust to anthracosis. Since all dust is primarily injurious on account of its irritating effects on the respiratory organs, it is evident that essential

[1] Ramazzini, *op. cit.*, pp. 183-184.
[2] Beddoes, Thomas, *Essay on the Causes, Early Signs, and Prevention of Pulmonary Consumption* (London, 1799), pp. 63-64.

variation in the mechanical properties of the dust must correspondingly affect the consequential results on lung tissue and the development of lung fibrosis and pulmonary tuberculosis.[1]   Of recent date some thought has been given to the possibility that there may be some colloidal action in the lungs and that some of the dust goes into solution.[2]   One important factor which is frequently overlooked in the analysis of dust hazards is the solubility or insolubility of the mineral dust inhaled, and the chemical nature of the dust may therefore be of even greater importance than its mechanically irritating qualities.   The degree of comminution is also of material importance, as in almost exact proportion to the degree of fineness will the dust particles penetrate into the remote portions of the lungs.   It is therefore held that the more minutely comminuted the dust, the more serious, in general terms, will be the damage to the respiratory organs.

Among the mineral dusts which are recognized as a distinct hazard is included buhrstone, which is one of the hardest stones known.   It is used for making millstones for the grinding of cereals and other substances.   According to Kober and Hayhurst, few buhrstone workmen live beyond the age of 35-40 years.   The workers complain that in the operation of chiselling the dust gets into the back of the throat and creates an exasperating sense of dryness which it is difficult to remove.[3]

The increasing use of Portland cement has created certain health problems in this industry, which, if not controlled, will adversely affect the health of large groups of workers.   The first plant for the manufacture of Portland cement in the

[1] Hayhurst, E. R., *A Survey of Industrial Health Hazards in Ohio* (Columbus, 1915), p. 17.

[2] Hefferman, P. and Green, A. T., "The Method of Action of Silica Dust in the Lungs," *Jo. Ind. Hyg.*, vol. x, no. 10, Oct., 1928, p. 272.

[3] Kober and Hayhurst, *op. cit.*, p. 704.

United States was built in 1872. Since that time there has been an extraordinary growth in the industry. In view of the increasing popularity of cement for use in road building and in the construction of buildings, the phenomenal growth of the industry is expected to continue. In order to appraise the health hazard, a careful and intensive study was undertaken of dust conditions in one of the large cement plants by the U. S. Public Health Service. The workers in this particular plant were under observation for a period of three years. From the evidence obtained by examining the men and keeping records of disabling sickness among them, it was found that certain diseases of the respiratory system occurred at a relatively high frequency among the employees of the cement plant, but that these diseases did not tend to become chronic.[1] Diseases of the eyes, ears, impacted wax in ears, ulceration of the nose with perforation of the nasal septum, as well as eczema or cement itch, have also been reported as being quite common among such cement workers.[2]

Potters and workers in the manufacture of china and earthenware are similarly exposed to hazards from mineral dusts. Potters frequently develop a condition which has come to be known as "potters' asthma," which has been widely accepted as typical of the highly specialized conditions under which respiratory affections resulting from continuous dust exposure in the pottery industry are likely to occur. These groups of workers in the pottery and china industries are exposed to particles of clay and flint, and to lead which is used in the glazes and for coloring purposes.[3]

The making, blowing and engraving of glass occupies a

[1] *U. S. Public Health Bulletin no. 176,* "The Health of Workers in Dusty Trades; Health of Workers in Portland Cement Plant," 1928, pp. 135-136.

[2] Kober and Hayhurst, *op. cit.,* p. 169.

[3] Oliver, *op. cit.,* p. 387.

prominent place among unhealthy trades. Glass workers are subject to four principal forms of industrial hazards, namely, (1) exposure to high temperature; (2) mechanical and chemical dust irritation; (3) poisoning by certain metals such as lead oxid, zinc oxid, or arsenic, used for coloring, etc.; and (4) irritation of the eyes from excessive heat and light.[1] Many of the workers are engaged in dry grinding, sanding, filing, drilling, beveling and polishing, especially in the crystal or so-called cut-glass industry in which a putty powder of a high lead content is used.[2]

The glass industry as a whole is a varied one, including among others the manufacture of blown and pressed ware, of window glass and plate glass, and so-called crystal or cut-glass. The labor division of the trade includes numerous and well-defined occupations, each of which is subject to more or less injurious circumstances, but of these the handling of materials and the mixing are the most liable to the risk of continuous inhalation of mineral dust. Glass blowers have always indicated a high death rate, but recent changes in manufacturing processes and the introduction of labor-saving machinery have resulted in sanitary improvements, which in consequence have had a favorable effect upon the health of glass blowers.

One of the outstanding hazards of mineral dust is that encountered in industries in which the workers are exposed to high concentrations of silica dust, resulting in a condition of the lungs known as silicosis. This may occur in factories in which abrasive stones are used; in the stone industry in which the use of pneumatic tools creates high concentrations of fine silica particles; in excavation, tunnel and subway work; and in mining. A number of reliable studies have been made of the silica-dust content in factories in which sil-

[1] Thompson, *op. cit.*, p. 467.
[2] Kober and Hayhurst, *op. cit.*, p. 174.

ica is liberated, and the findings have shown the definite health hazard to which the workers are exposed.[1] In the stone industry, studies made by Hoffman and others among the granite cutters of Vermont indicated a high death rate from tuberculosis among those engaged in this particular industry.[2] Various observers in different parts of the United States and Europe have noted the high sickness rate of the respiratory tract among stone and quarry workers exposed to the mineral dusts incidental to the preparation of marble, sandstone and granite for commercial and building use.[3]

The effect of mineral dust, such as silica, upon the health of those engaged in excavating, blasting and drilling rock of high silica content for building excavations, as well as for subways and tunnels, has been indicated in a recent study in New York City. A careful physical examination including X-rays of the lungs was made of a considerable number of workers engaged in these trades. In addition, special analyses were made of the dust which was created in the process of rock drilling and excavating in various locations in New York City. As a result of this particular study, it was found that 57 per cent of the men examined were suffering from silicosis in one stage or another, and that nine per cent

[1] Winslow, C. E. A. and Greenburg, L., " A Study of the Dust Hazard," *U. S. Public Health Repr.*, vol. 35, 1920.

[2] Hoffman, F. L., " The Problem of Dust Phthisis in the Granite Stone Industry," *U. S. Bur. Lab. Stat. Bull. no. 293*, May, 1922. For more recent study see: *U. S. Public Health Bull. no. 187*, 1929, " The Health of Workers in Dusty Trades "; " Exposure to Siliceous Dust " (Granite Industry), 206 pp. This report indicates a rapid rise in mortality from tuberculosis among granite workers since adoption of pneumatic hammer.

[3] Inter. Labour Office, *Occupation and Health*, Broch. No. 103, 1928, pp. 4-5; Kindel, A. J. and Hayhurst, E. R., " Stereoscopic X-ray Examination of Sandstone Quarry Workers," *Amer. Jour. Public Health*, vol. xvii, 1927, p. 818.

of the total number examined, all of whom were presumably well men regularly engaged in their occupations and working at the time of examination, had pulmonary tuberculosis in the active or inactive stage.[1] A study of a number of transcripts of death certificates of men who had been members of the rock drillers' union indicated that of a total of 21 natural deaths due to disease, 10 were due to pulmonary tuberculosis and seven to lobar penumonia. This was an abnormally high death rate from tuberculosis, as well as from respiratory diseases as a whole. These figures are in general agreement with findings among other workers engaged in drilling and blasting rock with a high silica content.

### MINERAL INDUSTRIES

Miners are subject to a variety of deleterious conditions which may be classed under: (1) those common to mining in general; and (2) those due to the specific substances mined. Among the harmful conditions due to mining in general may be mentioned the fact that with the increasing depth of the mine the temperature rises at the rate of approximately one degree F. per 100 feet of depth. With increasing depth, the difficulty of adequate ventilation becomes aggravated, and moisture as well as the heavier natural gases such as carbonic acid and marsh gas (carburetted hydrogen) are apt to accumulate. There are additional hazards to the health of the miners, including the gases generated in blasting with the use of such substances as nitroglycerin and dynamite; and various sanitary problems due to the inadequacy of proper facilities and soil pollution.

The methods of mining vary according to the nature of the ore mined, geological formations and other conditions.

---

[1] Smith, A. R., " Silicosis Among Rock Drillers, Blasters and Excavators in N. Y. City," *Jour. Ind. Hyg.*, vol. xi, no. 2, Feb., 1929, pp. 67-68.

Practically the entire mining industry is in a measure a branch of the stone industry, in that all metals and nonmetallic minerals secured by mining methods are obtained by processes of extraction from rock material more or less identical with the substances which enter into the stones quarried and cut for commercial purposes. Probably the outstanding industrial hazard to which miners are exposed is that of dust created in drilling, blasting and other dust-creating operations. The dusts attack the respiratory tract and give rise to different respiratory diseases, and more especially to some form of pneumoconiosis, which in turn may result in pulmonary tuberculosis.

There are nearly 100 different positions or specified employments in the coal-mining industry at present, including carpenters, machinists, electricians, etc. Most of the coal is mined in the proper sense of the term, underground, though within recent years some coal-stripping operations have been developed. Primitive methods of coal mining with pick and shovel have largely been replaced by coal-cutting machinery, which has apparently resulted in an increase in the dust contamination of the atmosphere. On account of the extensive use of such machinery the technical requirements for underground work are apparently less at the present time than in former years, so that the term "miner," with special reference to coal, has lost much of its former restricted significance.

Coal miners have been found to be less susceptible to pulmonary tuberculosis than other miners, this being due in large measure to the chemical composition of the coal.[1] However, long exposure to the fine particles of coal may result in a lung condition known as anthracosis, which is a type of fibrosis of the lungs. A high frequency of asthma and

[1] *U. S. Bur. Labor Stat. Bull. no. 231*, p. 381.

bronchitis has been noted among such miners, particularly in Great Britain. Due to more hygienic methods of mining, as well as to more improved methods of ventilation, there has been a notable decrease in the respiratory conditions heretofore so common among coal miners.[1]

Among other substances which are mined may be mentioned asphalt, bauxite, copper, gold, silver, graphite, iron, lead, zinc, mica, phosphate, quicksilver and sulphur. In mining these various substances, the severity and character of resulting ailments and diseases depend in large measure upon the chemical composition of the dust generated, as well as upon the dust concentration, and the length of time to which the worker is exposed to the particular kind of dust. In a careful study embodied in the report to the Secretary of State for the Home Department in 1904 on the health of the Cornish miners, J. S. Haldane and his co-workers took occasion to point out the serious health hazards to which these workers were being exposed. The report indicated that the death rate among the miners living in Cornwall, where tin was mined, was very high; that the excessive death rate was due entirely to phthisis and other lung diseases; that the excessive deaths were to be found among those who used the pneumatic rock drills; and that among the men examined at least two-thirds of them were found to be suffering from tuberculosis due to the inhalation of silica dust.[2] In this country, A. J. Lanza with the collaboration of E. Higgins, published his first study of the subject in 1915 for the U. S.

[1] *Ibid.*, p. 380. Statistical evidence in England indicates that men exposed to coal dust seem to succumb in excess from bronchitis and pneumonia, but not from phthisis. See, Collis, E. L. and Gilchrist, J. C., "Effects of Dust upon Coal Trimmers," *Jour. Ind. Hyg.*, vol. x, no. 4, April, 1928, p. 109.

[2] Haldane, J. S., Martin, J. S. and Thomas, R. A., *Report on Health of Cornish Miners* (London, 1904), pp. 20-21.

Bureau of Mines.[1] This was followed by other studies on the same subject of respiratory diseases in the mining industry, these later reports being published in 1917[2] and 1921.[3]

Some outstanding studies have been made by the Union of South Africa Miners' Phthisis Medical Bureau on the prevalence of miners' phthisis and silicosis among the European and native gold miners.[4] This organization is carrying on its studies and the periodic reports of the findings indicate a continued hazard to which the workers are being exposed, though much has been done since the organization of the Bureau to remedy conditions through ventilation, periodic examination of the workers, and other possible measures.[5]

As a result of the measures for control which have been instituted, a further drop in the cases of simple silicosis occured in 1927-1928 and in 1928-1929. This has been held to be most significant, inasmuch as the decrease in cases occurred in the face of a continued increase in the number of older miners employed.[6]

In gold and silver mining, as in the case of some other

[1] Lanza, A. J. and Higgins, E., "Pulmonary Diseases among Miners in the Joplin District," *U. S. Bur. Mines. Technical Paper no. 105*, 1915.

[2] Higgins, E., Lanza, A. J., Laney, F. B., Rice, G. S., "Silicosis Dust in Relation to Pulmonary Diseases among Miners in the Joplin District, Missouri," *U. S. Bureau Mines, Bull. no. 132*, 1917; see also, Lanza, A. J. and Childs, S. B., "Miners' Consumption," *U. S. Public Health Bull. no. 85*, Jan., 1917.

[3] Harrington, D. and Lanza, A. J., "Miners' Consumption in the Mines of Butte, Mont.," *U. S. Bur. Mines Tech. Paper*, no. 260, 1921.

[4] *Report of Miners' Phthisis Medical Bureau, Pretoria* (Union of So. Africa), 1929, contains comparative figures of incidence of tuberculosis between the years 1917-1928 inclusive, pp. 30-35.

[5] Watkins-Pitchford, W., "The Silicosis of the South African Gold Mines, and the Changes Produced in It by Legislative and Administrative Control." *Jour. Ind. Hyg.*, vol. ix, no. 4, April, 1927, pp. 109-139.

[6] *Report of Miners' Phthisis Medical Bureau*, 1930, p. 19.

metals, extracted by the ordinary processes of deep mining, the metal itself constitutes but a minute fraction of the ore, which, however, may frequently contain a relatively large proportion of lead and other minerals. The rock itself is generally a silicious quartz, which, during the process of rock drilling, seriously contaminates the underground atmosphere, often with disastrous results to the lungs of the miners. The term " gold and silver mining ", is, therefore of importance only insofar that mining for the purpose of extracting the precious metals is, as a general rule, inseparable from exposure to more or less health-injurious silica dust. Many gold and silver mines, as measured by the metallic content of the ore, are more accurately described as lead mines.[1]

Throughout this country as well as in foreign lands, the underground conditions vary considerably on account of important differences in the chemical and mechanical properties of the rock dust inhaled during drilling and other underground operations. As has been noted, careful studies of the injurious effects of the dust created in mining operations have only been initiated within recent years, though earlier writers already mentioned were aware of the harmful effects of mining in general.

Akin to the health-injurious conditions in gold and silver mines with particular reference to pulmonary tuberculosis and non-tuberculous respiratory disease, are those in the copper mining industry. It is not so much the copper itself which causes the respiratory diseases as the rock dust in fine particles which is taken into the lungs of the mine workers, and after some years of exposure, develops some respiratory disease. As in the case of other metals, copper represents but a small amount of the total ore extracted.[2]

Among other miners who are seriously affected by contin-

---

[1] *U. S. Bur. Labor Stat. Bull. no. 231*, p. 352.

[2] *Ibid.*, p. 366.

ued exposure to dust in their occupations, are those employed in lead and zinc mines. At the present time a careful study of the effect upon the health of such miners of the dust to which they are exposed is being conducted by the U. S. Bureau of Mines and the Metropolitan Life Insurance Company at Picher, Oklahoma.[1] No reports are as yet available, though it is unquestionable that the hard rock dust is having a bad effect upon the lungs of the miners.

The main hazards to which miners are exposed result from the specific substances mined, as well as from the large quantities of fine silica dust which enter the lungs of the miners. Various remedies have been proposed, including adequate ventilation, proper spacing of miners, shortening of the working day, prohibition of dry drilling, etc. All of these measures and others that have been tried have not as yet solved the problem of safeguarding the health of mine workers.

ORGANIC AND MISCELLANEOUS DUSTS

### Vegetable Fiber Dust

The number of industries and occupations in which the worker is exposed to vegetable fiber dust is rather large. Among the more important ones may be mentioned the following: broom and brush factories; cotton ginners; cotton spinners and weavers, as well as other cotton-mill employees; hay and straw bailers; hemp and jute mills; knitting mills; lace and embroidery workers; linen mills; paper and pulp mills; rope and cordage factories; textile mill workers; cabinetmakers and woodcarvers; workers in furniture factories and other woodworkers and handlers. In dealing with the

[1] This study is in part a continuation of the work of the U S. Bureau of Mines at Joplin, Mo., which was first undertaken in 1915; see, *Amer. Jour. Public Health*, vol. xix for details of clinic work, June, 1929, pp. 635-640.

hazards of vegetable-fibre dust, the same conditions are usually with as in other dust-producing industries. The dust is generally a definite causative factor in the development of respiratory diseases, the severity of the disease usually being determined by the kind of dust, fineness of the particles, and length of time the worker is exposed to the dust. In addition, the dust concentration in the atmosphere of the workplace, as well as the humidity and ventilation adequacy play an important part. As the definite hazards in a number of the above-mentioned industries and occupations overlap, mention will only be made of such as are representative of definite groups.

The work of cotton-mill employees involves more or less constant confinement in a dusty atmosphere, even in the best regulated mills. The intrinsic danger in the industry in this respect lies chiefly in the opening, picking and carding processes, the danger varying with the construction of the mill, the amount of dirt and other impurities in the stock, the means of removing the dust, and certain other factors. Conditions commonly met with in weaving and spinning rooms are: poor light; presence of carbon dioxide and carbon monoxide in the air; non-regulation or unscientific and unsatisfactory regulation of artificial moisture; excess of moisture; undue heat; and want of provision for a plentiful supply of fresh air.[1] Raw cotton contains much dirt composed of earth and sand, and in the ginning process the dirt, cottonseed fragments and cotton fibers are blown about so that the air is filled with irritant dust. More or less respiratory irritation results. Nose bleed and nasal catarrh are common and the workers complain of the dryness of the nose and throat.[2]

The dust hazard exists in addition to the frequently poor

[1] Kober and Hayhurst, *op. cit.*, pp. 257-259.
[2] Thompson, *op. cit.*, p. 423.

working conditions in many of the mills, particularly in the smaller establishments. Cotton spinners have to work all the year round in a very warm and humid atmosphere, and accordingly suffer from debility and exhaustion caused by profuse sweating. The temperature and moisture are maintained at a high standard both day and night in order to prevent brittleness in the cotton fibers, and as a consequence the workers became peculiarly sensitive to chills, brought about probably by injudicious exposure to draughts.[1]

The dust from flax and hemp act in substantially the same manner. Being volatile, these dusts are easily inhaled, and in addition, they are often mixed with earth, oil from machinery used in the plants, and other substances. The process of beating out the fibers liberates a great deal of dust composed of sharp-pointed plant cells, dirt, etc. Automatic machinery is gradually replacing the dusty hand methods of breaking and binding the fibers.[2] No recent precise data exists as to the frequence of tuberculosis and other respiratory diseases among flax and linen workers. There is general agreement that tuberculosis is fairly high among the workers engaged in this industry. Linen workers are also exposed to mill fever, an affection which generally lasts three or four days, and, as in the case of other textile workers, is probably due to the phenomenon of anaphylaxis. The asthmatic attacks these workers suffer from are undoubtedly due to the inhalation of dust in the place of employment, while the morbidity rate is seemingly very high.[3]

From the point of view of the health of the workers, the handling of hemp offers in a rather more accentuated degree all the disadvantages that arise from the working of flax. There are, besides, some extra disadvantages. The workers

[1] Oliver, *op. cit.*, p. 147.

[2] Thompson, *op. cit.*, p. 427.

[3] Inter. Labour Office, *Occupation and Health*, Broch. no. 96, 1928, p. 6.

suffer from a whole series of lesions. First and foremost comes a very general erythema or skin rash, which attacks the workers who pull off the leaves; it is due to the dust, and is aggravated by the perspiration and by the physical exertion of the work. The legs and feet of the workers become affected with eczema, which is probably of traumatic origin and may easily be complicated by deep and painful wounds, by irritation, etc. Hemp dust, which is thick and made up of rather coarse and irregular as well as pointed particles, causes irritation of the conjunctiva and the mucous membrane of the respiratory passages—the nose, back of the throat and pharynx.[1]

The paper industry subjects the workers to several occupational hazards and diseases. Paper is made from a variety of substances in this country, such as from old rags, old paper, wood pulp and burlap. In the thrashing and chopping of the old rags much dust is stirred up, resulting in various respiratory affections. Chronic bronchitis, emphysema, granular inflammation of the eyelids, and impacted ear wax are fairly common among the paper workers. There also exists the danger from the use of chloride of lime for bleeching purposes, as well as the use of toxic colors containing lead or arsenic. In the production of paper from wood pulp by the sulphite cellulose process there is the hazard due to the escape of sulphur dioxide from the sulphur stoves or from the boilers.[2]

Ordinary wood dust has attracted attention to its injuriousness to the worker by reason of its purely mechanical action, though the handling of certain kinds of woods, especially such as come from some foreign lands and the tropics, requires still greater care because of the essential oils impregnating them which, when liberated in the dust incident to

[1] *Ibid.*, no. 99, 1928, p. 4.
[2] Kober and Hayhurst, *op. cit.*, p. 271.

manufacturing processes, may affect the health of the workers concerned. Construction timber of Europe and other temperate regions sometimes comes from trees, such as the acacia, sweet chestnut, yew and juniper, which contain poisonous or injurious substances, but they are generally innocuous or contain such minute quantities of the toxic agent that very little trouble is attributable to them. On the other hand, quite a number of woods from tropical countries, much sought after because of their special qualities for fine cabinet making and for use in the dyeing and perfume industries cause, according to their genus, either an irritation of the skin, mucous membrane, or symptoms of general poisoning.[1] Some years ago it was noticed in England that various toxic and other symptoms were occurring with considerable frequency among those employed in the manufacture of wood products from African boxwood. Investigation revealed the presence of an alkaloid in the wood which acted as a heart depressant and produced a number of other symptoms indicating exposure to some toxic substance.[2]

Apart from the dust hazard, woodworkers such as furniture makers, carpenters and cabinetmakers, are liable to suffer from callosities and contracted tendons of the palm of the hand. Others develop occupational neuroses and peculiar inflammation of the tendons of the thumb, frequently resulting in incapacity to pursue the particular occupation for which the workers have been trained.[3]

### ANIMAL AND MIXED FIBER DUST

In general, it may be stated that animal and mixed fiber dust exposure is to be found in hat factories; among furriers and hair workers; in carpet mills; silk mills; woolen and

---

[1] Inter. Labour Office, *Occupation and Health*, Broch. no. 22, 1925, p. 1.
[2] Kober and Hayhurst, *op. cit.*, p. 204.
[3] *Ibid.*, p. 204.

worsted mills; and among mattress makers and upholsterers. In the felt hat industry the animal hairs are subjected to special treatment for felting purposes by treatment with a mercurial solution. Despite its highly poisonous character, mercury continues to be used for this purpose. Efforts have been made to substitute a non-poisonous metal for mercury but they have not thus far met with success. As the matter of mercury poisoning will be dealt with at length in connection with the discussion of other metal poisons, only the dust hazard of the felt hat industry will be gone into at this point.

In the preparation of hat bodies, the rabbit, hare and nutria skins are opened, shaken out and brushed, in order to remove dirt and loose hair; the long hair are then removed by plucking, leaving the short soft hairs, called the fur. In this and other processes which follow in the preparation of the finished felt hat considerable dust arises and irritates the respiratory tract of the worker.[1] In a study conducted in Danbury, Conn., in which 108 hatters were carefully examined, respiratory infections were reported as common. Forty-nine individuals gave a history of different colds, 16 of pneumonia, 18 of pleurisy, 49 of frequent coughs and 9 of hemoptysis. Examination of the lungs disclosed various positive findings in 32 cases, for the most part signs of bronchitis or of old pleuritis. X-rays of the lungs indicated further defects, though none of these workers were found to have pulmonary tuberculosis at the time of examination. It should of course be noted that the men examined were regularly at work, with the probability that those suffering from an active case of tuberculosis in an aggravated form

[1] Johnson, J. R., *The Preparation of Hatters' Fur. Jour. Ind. Hyg.,* Dec., 1922, vol. iv, no. 8, p. 325. For experience in England and Wales, see, *The Registrar-General's Decennial Supplement for England and Wales,* 1921, pt. ii, p. lxxix.

could not find it possible to continue at such a trade for any length of time.[1]

Furriers are similarly exposed to most of the hazards which afflict the felt hat workers, though, instead of mercury poisoning, furriers are exposed to various aniline dyes. In a careful examination of 889 persons employed in the fur industry in New York City, it was found that there were definite occupational hazards to which the workers engaged in the various processes were liable. Asthma, bronchitis and pulmonary tuberculosis were noted in this study.

Skin lesions of many kinds were found among these workers in such proportion as to indicate a definite health hazard in the industry. Nose and throat affections were frequently encountered, a number of furriers showing the presence of dye pigments and sawdust in their nostrils. The various defects discovered are largely due to the dust peculiar to the character of the industry and the materials handled.[2]

Workers in carpet and woolen and worsted mills are similarly exposed to dust arising from the handling of the wool entering into the manufactured product. The sorting and cleaning of wool is a very dusty occupation, for wool is oily and contains considerable extraneous dirt. Wool-sorters often suffer from both acute and chronic catarrh of the air passages. Various chemicals are used in the many processes of handling the raw product and preparing it for use in the manufacture of carpets and cloths.[3] Mattress makers and upholsterers are exposed to the fine particles of dust arising from the handling of horsehair. Hair dust irritates the

---

[1] Wright, Wade, "A Clinical Study of Fur Cutters and Felt Hat Workers," *Jour. Ind. Hyg.*, vol. i, no. 7, Nov., 1922, pp. 296-298.

[2] Harris, L. I., "A Clinical and Sanitary Study of the Fur Hatters' Fur Trade," *N. Y. City Dept. of Health Monograph no. 12*, Dec., 1915, pp. 10-14.

[3] Thompson, *op. cit.*, p. 448.

nasal mucosa and gives rise to sneezing, ulceration and catarrh, as well as to asthmatic attacks.

### ORGANIC DUST

Following the classification suggested by Hoffman, it might be noted that among industries in which organic dust is to be found in sufficient concentration to possibly affect the health of the workers, the following might be included: bakeries; factories producing objects of bone and ivory, celluloid and rubber; factories for the manufacture of candy, cigars and tobacco products; fertilizers, gloves, harness, pocketbooks and other leather objects, such as shoes; charcoal and coke works; grain and flour mills; and tanneries. A discussion of a few of these industries will indicate some of the health hazards to which the worker is exposed in the daily routine of his work.

The grinding of various materials such as horn, hoofs, bone, ivory, vegetable ivory, tortoise shell, celluloid, sea shells, etc. of which buttons and other objects are made, distributes clouds of dust. The emery grinding wheels used for turning and polishing the buttons and similar objects help to scatter the dust throughout the workplaces. Mortality figures among button makers from pulmonary tuberculosis indicate a rate more than twice that in the registration area for workers of a similar age distribution.[1] Such workers acquire rhinitis, chronic bronchitis, conjunctivitis, and may develop fibroid phthisis. In addition, there is considerable asthma among button makers.

Among bakers, the lodgment of flour and sugar dust upon the teeth and gums unless removed by frequent cleaning, favors fermentation processes and decay of the teeth. Other factors prejudicial to the health of bakers are exposure to heat, coal gas, impure air, and handling of the flour in the

[1] Inter. Labour Office, *Occupation and Health*, Broch. no. 78, 1927, p. 1.

form of dough resulting in eczema of the hands and arms, sometimes known as "bakers' itch". They are particularly susceptible to diseases of the respiratory tract, due to irritation from the flour, the exposure to extreme temperatures and rapid cooling of the body, excessive perspiration and insanitary work places.[1]

There has been considerable discussion for many years as to whether or not the tobacco industry was injurious to the health of the worker. In this industry there are various processes involved in the preparation of the tobacco leaf for use in the manufacture of cigars, cigarettes, chewing tobacco and snuff. The dust created in handling the tobacco leaf in the various processes may be considerable in plants in which the work is all done by hand, while it may be reduced to some extent where the processes are in part at least carried out by the use of machinery. However, in spite of the improvements and the better ventilation in up-to-date factories, the workers are still obliged to breathe an atmosphere containing both dust from the tobacco and vapors from the volatile oils which give the tobacco leaf its special perfume. It is very likely that the workers, especially the younger ones, suffer sooner or later from their unwholesome occupation.

Frederick L. Hoffman has shown that the United States statistics for 1909 gave cigar makers and tobacco workers a higher death rate for all causes in each age group than they did the manufacturing and mechanical class or the mercantile and trading class. He concluded that tobacco workers were subject to an excessive mortality from pulmonary tuberculosis and from other respiratory diseases.[2] The Metropolitan Life Insurance Co. conducted a study of deaths among cigarmakers and tobacco workers who were industrial policy holders of the company during the years 1911 to 1913

[1] *Ibid.*, no. 88, 1928, p. 6.

[2] Hoffman, F. L., *U. S. Bureau Labor Stat. Bull. no. 82*, 1909, p. 569.

inclusive. Pulmonary tuberculosis appeared as the leading cause of death among these workers; in the age period 15 to 24 years the proportionate mortality from tuberculosis was responsible for 48.5 per cent of the total deaths, as compared with 33.8 per cent for all occupations.[1] In a similar study covering the years 1922-1924 a high death rate from pulmonary tuberculosis was also found.[2] Official figures issued in 1913 by the state tobacco factory in Baden, Germany, indicated that the death rate from pulmonary tuberculosis among the workmen was 5.7 per cent against 1.9 per cent among the rest of the inhabitants.[3] An analysis of the records of the Department of Health of San Juan and Rio Piedras showed that of the workers who were examined by the physicians of the Department, the following percentages of tuberculosis were discovered; incidence of tuberculosis among food handlers 0.9 per cent; incidence of tuberculosis for the entire group examined 1.1 per cent; incidence among workers in the tobacco trade 1.2 per cent.[4]

In 1917 the Industrial Board of the Pennsylvania State Department of Labor and Industry asked the clinic for diseases of occupation of the University Hospital in Philadelphia to study the health hazards in cigar factories. A study was made of 400 workers. While only a comparatively small number of the workers were found to be suffering from either inactive or active pulmonary tuberculosis, it was presumed that many of those in whom only certain vague physical signs were found might eventually die of tuberculosis; otherwise there was no apparent explanation of the

[1] *U. S. Bur. Labor Stat. Bull. no. 207,* 1917, p. 25.

[2] *Ibid., no. 507,* 1930, p. 27.

[3] Schrumpf, P., *Tobacco and Physical Efficiency* (New York, 1927), p. 44.

[4] Nazario, Ramon, C. R., " Epidemiology of Tuberculosis in Porto Rico," *Porto Rico Rev. of Public Health,* vol. iv, no. 4, Oct., 1928, p. 179.

known high death rate from tuberculosis among cigar and tobacco workers and the few cases discovered among the 400 workers examined.[1]

There have been varying opinions expressed by competent observers on the health injuriousness of the tobacco industry. Some have held that the industry is not injurious to the workers, whereas others equally competent have taken the opposite view. Despite the difference of opinion, which is based largely on the observation of active cases of pulmonary tuberculosis or the seeming lack of such cases in the tobacco industry, the certified deaths to an excessive degree from tuberculosis among such workers indicate that there is some definite basis for the position that the industry is hazardous. A further explanation might lie in the fact that possibly many who are not in robust health enter upon this work, and that the constrained position of the workers and the generally poor atmosphere of the work places may predispose to tuberculosis and other diseases of the respiratory tract. Also, there may be a racial selection of employees, with Porto Ricans and others with known high death rates from tuberculosis, forming a considerable proportion of the workers.[2]

Certain outdoor workers are exposed to mixed organic and inorganic dusts. These workers include chauffeurs, drivers and teamsters, garbagemen, street car conductors and motormen, street cleaners, etc. The dust concentration is not sufficient to cause any serious occupational hazard, while at the same time the hours in the open offset in part at least any serious danger to which these workers might be subjected.

[1] Miller, T. G., "A Sociologic and Medical Study of Four Hundred Cigar Workers in Philadelphia," *Amer. Jour. Med. Sciences* (Philadelphia), vol. 155, Feb., 1918, pp. 157-173.

[2] Kober and Hayhurst, *op. cit.*, p. 227.

### TUBERCULOSIS AND DUST

The inhalation of dust has long been recognized as being injurious to the health of the workers. However, when pulmonary tuberculosis was not as yet clearly distinguished from other respiratory diseases, the precise disease alluded to in earlier writings on occupational diseases cannot always be definitely recognized. Thus, Simmons, writing in 1780 on the *Practical Observations on the Treatment of Consumption,* noted that: " Scythe grinders from breathing the dust of the sandstone and iron, are affected with a consumption which they denominate the grinders' rot." [1] Certain events in the 19th century came to help clear up the situation. Pulmonary tuberculosis, or phthisis, as it was called, was more definitely recognized as a clinical entity; mortality statistics were collected, and the tubercle bacillus was discovered. Medical opinion up to the beginning of the 20th century continued to hold as a broad truth that the inhalation of any kind of dust predisposed to diseases of the lungs, of which pulmonary tuberculosis was the chief. The intensive investigations carried on by A. L. Collis and others into the prevalence of pulmonary tuberculosis in certain occupations and the amount and kind of dust present, indicated that at any rate so far as pulmonary tuberculosis was concerned, the kind of dust was of primary importance.[2] It is generally recognized that tuberculosis is still the industrial disease of greatest importance in spite of its rapid decrease in the general population during recent years.

### CONCLUSION

The administrative control of tuberculosis in American states and cities has heretofore been concentrated chiefly

[1] Simmons, S. F., quoted in Thomas Young, *Historical Treatise on Consumptive Diseases* (London, 1815), p. 290.

[2] Inter. Labour Office, *Occupation and Health,* Broch. no. 62, 1926, p. 7.

upon legislative enactments and their enforcement for the registration of existing cases; the voluntary or compulsory segregation and detention of tuberculous patients; the establishment of federal, state, county and municipal tuberculosis sanatoria; and the enforcement of sanitary ordinances against indiscriminate expectoration in public places. Some progress has been made in the direction of labor legislation aiming at the control of tuberculosis in industry, principally in the dusty trades, but the results thus far have not been satisfactory. This has been so chiefly because of an inadequate realization of the seriousness of the problem. The statistical evidence that certain trades or occupations are distinctly more unfavorable to health and longevity than others is quite conclusive. The regulation of industry with special reference to the dusty trades and tuberculosis is a national and state labor problem of the first order in practical importance.

While tuberculosis stands out as perhaps the most serious condition to which workers in dusty trades are exposed, there are in addition, other respiratory diseases which are directly traceable to the dust exposure involved in some industries. To focus attention upon these hazards will eventually lead to studied attempts to control the dust. In addition, with the increase in information relating to these particular occupations and industries, it is safe to presume that efforts will be made to provide adequately for the worker and his family when constant or even intermittent exposure to some industrial dust has seriously affected the health, working capacity and life expectation of the worker.

# CHAPTER III

## Hazards in Metal, Chemical and Miscellaneous Industries

### INTRODUCTION

For some years the question of unhealthy industries has attracted the attention of experts and others interested in the health and welfare of the worker. There has been an increasing demand for legislative action likely to reduce to a minimum the industrial dangers to which the workers in certain industries are exposed. Although certain industries cannot for the time being be made entirely innocuous, the progress of modern science and technique should tend to make some industrial processes less dangerous than they are at present. On the other hand, scientific achievement leads to the discovery or creation of products and processes which, when adopted in industry, are a source of new dangers to the workers. Unfortunately, occupational risks do not lurk only in the raw materials, in the finished products, or in by-products, but are often concealed in intermediate manufacturing operations, and are only detected by experts when disease has already attacked the worker.

From the point of view of the wage-earner there is probably no material difference insofar as end results are concerned, as to whether the industrial hazard to which he has been exposed, and as a result of which he has become incapacitated, arose from dust, or lead, or other irritant and poison. To him the consequences are just as serious.

60

There is a vast difference, however, to the physician, the engineer or other scientist. If progress in aiding and protecting the worker is to be achieved, it will come through the efforts of scientists in determining the causes, in providing measures for the prevention and regulation of the hazards, and in developing methods of treating the particular ailments. These desirable ends can only be achieved by persistent study of the different hazards to which the worker is exposed.

The metal industries have long been recognized as contributing an undue proportion of cases of occupational disability. Due to the very extensive use of metals in the arts, crafts and industries, many workers are daily exposed to the deleterious effects of metals in both the pure and alloy states, with various physical disabilities and defects arising in the workers handling and working with them. Of the many metals and compounds used, the following may be classified as irritants which often lead to adverse effects upon the health of the workers: antimony, arsenic, brass, bronze, chrome pigments, cobalt, copper, gold, iron, steel, lead, manganese, mercury, nickel, platinum, tin, vanadium and zinc.

### LEAD AS AN INDUSTRIAL HAZARD

Of the hundreds of materials causing industrial disease, lead is probably the most destructive. It is not as virulent as other substances, but, on account of its wide use, causes more loss of time due to illness and death than all other occupational poisons. Either as a metal, a salt of the metal, a paint, a solder, or occurring as a droplet, dust or fume, it does its damage, but usually in a slow, insidious way.[1] The metal itself is comparatively harmless; the dust is haz-

[1] Hackett, J. D., *Health Maintenance in Industry* (New York, 1925), p. 163.

ardous according to fineness, and the fume is perhaps, more harmful than the other forms. Lead and its compounds vary in poisonousness; the soluble is more poisonous than the insoluble; the oxides more than the sulphates; the volatile forms more than the non-volatile.

The use of lead for various industrial processes and for painting was well known to the ancients. Pliny speaks of white lead, and a method of corroding lead as a means by which white lead was made for paint.[1] Of all the metals employed in the arts and industries, none lends itself to such applicability as lead. In its metallic state it is so plastic that it can be readily moulded. It forms compounds which for color and persistence have enduring properties superior to most of the metals. There are few articles of manufacture that have not been directly or indirectly brought into contact with lead, and in many of the newer industries the association is very close.[2]

Lead may get into the human system by the inhalation and ingestion of lead-containing dusts, paints, pastes, enamels or oils; by the inhalation and ingestion of lead-containing fumes from the molten processes; and by absorption through the skin, though this is generally questioned at present.[3]

A study conducted in New York has indicated the relative lead hazard in the handling of various lead substances. The following table indicates the percentage of cases of lead absorption.[4]

[1] Legge, T. M. and Goadby, K. W., *Lead Poisoning and Lead Absorption* (London, 1912), p. 1.

[2] Oliver, Sir Thomas, *Dangerous Trades* (London, 1902), p. 282.

[3] "Survey of Industrial Health Hazards and Occupational Diseases in Ohio," *Ohio State Board of Health*, 1915, p. 378.

[4] *Annual Report*, Industrial Commissioner, N. Y. State Dept. of Labor, 1927, p. 200.

## TABLE I

### Lead Hazard in Handling of Lead Substances

| Process or Materials Handled | No. of Cases | Cases of Lead Absorption | Per cent |
|---|---|---|---|
| 1. Lead burning (lead fumes) ......... | 10 | 10 | 100.0 |
| 2. Handling metallic lead (cold) ...... | 66 | 49 | 74.2 |
| 3. Handling dry pottery glaze ......... | 19 | 14 | 73.6 |
| 4. Mixing dry chrome colors .......... | 11 | 8 | 72.7 |
| 5. Pottery glaze dippers .............. | 14 | 10 | 71.4 |
| 6. Handling red lead powder .......... | 63 | 41 | 65.0 |
| 7. Handling red lead paste ............ | 23 | 15 | 64.7 |
| 8. Handling wet paint ................ | 13 | 8 | 61.5 |
| 9. Mixing dry ingredients of paint ..... | 10 | 6 | 60.0 |
| 10. Handling molten lead (lead pots) ... | 43 | 25 | 58.1 |
| 11. Handling white lead powder ........ | 26 | 15 | 57.6 |
| 12. Handling metallic lead (hot) ....... | 16 | 9 | 56.2 |
| 13. Lead smelting .................... | 33 | 18 | 54.5 |
| 14. Spraying paint ................... | 9 | 4 | 44.4 |

In considering the problem of lead poisoning, it should be noted that susceptibility and immunity to lead poisoning are influenced by certain factors, such as age and sex. It is generally accepted that young persons are more liable to lead poisoning than adults and women are more susceptible than men.[1]

Hoffman[2] studied a total of 1,592 original death certificates representing the recorded mortality from lead poisoning in the United States registration area for the years 1914-1924 inclusive. The deaths represented among others the following; 62 lead workers; 35 metal miners; 841 painters; 25 plumbers; 67 printers; and 48 in miscellaneous occupations. The average age of those dying from chronic lead poisoning was 49.3 years; lead workers averaged 45.3 years; painters 51.3 years; plumbers 48.2 years; printers 45.4

[1] Legge and Goadby, *op. cit.*, p. 35.

[2] Hoffman, F. L., "Deaths from Lead Poisoning," *U. S. Bureau of Labor Statistics, Bull. no. 426*, Feb., 1927, p. 16. See: *U. S. Bur. Labor Stat. Bull. no. 488*, June, 1929, reporting a further decline in deaths.

years; while carpenters and joiners averaged 57.5 years. Though it is recognized that many factors are involved in influencing the length of life of an individual, it is nevertheless significant to note that those exposed to the largest dosage of lead have the lowest length of life.[1]

White lead (carbonate of lead) is a fine white powder, and is in considerable demand for making paint, for plumbing, and for making the glaze for pottery. The white lead industry, which formerly was a prolific cause of lead poisoning, has undergone great improvement in recent years, not only in regard to construction and operation, but also in the personal care of the workers. Lead poisoning cannot as yet be regarded as anything but a very present danger in a white lead plant.[2]

In England the manufacture of red lead (monoxide of lead) has been regarded by some physicians as only slightly dangerous to the health of the workmen. Oliver found that such was not the case, and reported several instances of plumbism due to the inhalation of red lead dust.[3] In the United States red lead has similarly been found to be harmful, and in instances where the manufacture of the oxide of lead was carried on without due precautions, it was found that there were more risks entailed by the worker than in white lead plants.[4]

The dangers in the making of white lead are in the fumes from the furnaces, and in the dust from dumping, grinding, screening and packing the oxides. In some factories mech-

[1] For English experience in deaths from lead poisoning see: *Registrar-General's Decennial Supplement for England and Wales*, 1921, pt. ii, p. li.

[2] Hamilton, Alice, *Industrial Poisons in the United States* (New York, 1925), p. 165.

[3] Oliver, Sir Thomas, *Lead Poisoning* (London, 1914), p. 19.

[4] International Labour Office, *Occupation and Health Brochure no. 14* (Geneva, 1925), p. 2.

anical furnaces, well hooded, have done away with fumes, and other mechanical processes have eliminated much of the dust.   In addition, the employment of physicians to examine and treat the men has similarly helped.   Still, the falling off in the sickness rate cannot be proven statistically, since the plant physicians if they report cases of lead poisoning at all, usually report only the severer ones.[1]

## PAINTERS

At the present time the painters' trade is regarded in all countries as the most important of the lead industries, for it employs a large number of men and, as it is a skilled and comparatively well-paid trade, men generally do not drop out of it at the first attack of illness, but usually keep on till actually incapacitated.[2]   It is harder to control conditions in the painting trade than in any other lead trade.[3]

The dangers which the painter who uses lead paints must face are generally as follows: (1) mixing dry lead salts with oil or paint; (2) sandpapering lead-painted surfaces; (3) rubbing or chipping off old paint; (4) burning off old paint; (5) inhaling dust from dirty working clothes and from dirty drop cloths; and (6) carrying lead paint into the mouth from unwashed hands while eating or while handling tobacco.[4]

[1] Hamilton, Alice, *Lead Poisoning in the United States,* in Kober, G. M. and Hayhurst, E. R., *Industrial Health,* 1924, p. 445.   For influence of medical departments on reduction of lead poisoning see: *U. S. Bur. Labor Stat. Bull. no. 120,* May, 1913, p. 22.

[2] For lead poisoning among painters in former years, see: Ramazzini, *op. cit.,* p. 72.

[3] Hamilton, Alice, *Industrial Poisons in the United States,* p. 187; see *Monthly Labor Review,* vol. ix, no. 1, July, 1919, p. 172.   For an extensive bibliography on lead poisoning, including 500 titles, see: *Lead Poisoning,* by Aub, J. C., Minot, A. S., Fairhall, L. T., Reznikoff, P. (Baltimore, 1926), 265 pp.

[4] *U. S. Bur. Labor Stat. Bull. no. 120,* 1913, " Hygiene of the Painters' Trade," p. 33.

Sandpapering paint and burning off old paint are not dangers to which every painter is exposed. To eliminate some of the hazard, the dry sandpapering of lead paint has been prohibited by law in several countries.[1] It is difficult to state how much of the lead poisoning of painters is caused by the lead paint which is so often smeared over their hands. Some have assumed that skin absorption is practically negligible as a mode of industrial poisoning,[2] but that belief rests on an assumption that has never been proved.[3]

Out of the total of 1,592 deaths from chronic lead poisoning in the United States during the period of 1914-1924 inclusive, 841, or 52.8 per cent were deaths of painters, who died at an average of 51.3 years. Hoffman found a trend which he believed to indicate that there was a decline in lead poisoning in the United States, at least insofar as painters are concerned. By compiling these deaths for single years, it appeared that there were 407 deaths of painters during the five years ending with 1915, against 360 deaths during the five years ending with 1923, while there was an increase in the average age of death from 50.6 years during the first five years to 51.9 during the last period. In the year 1925, there were 74 deaths of painters, at an average age of 52.6 years.[4]

In 1902 the French Government, by a decree applying to house painting, prohibited (1) use of white lead except

---

[1] Kober and Hayhurst, *Industrial Health*, p. 454.

[2] Mayers, M. R., "Lead Absorption," *Industrial Hygiene Bull. N. Y. State Dept. of Labor*, vol. v, no. 6, Dec., 1928, p. 1.

[3] Hamilton, *op. cit.*, p. 191.

[4] Hoffman, F. L., "The Decline in Lead Poisoning," address before Health Congress of the Royal Institute of Public Health, Belgium, June 1, 1927. For more recent figures on decline, see Hoffman, "Deaths from Lead Poisoning, 1925-1927," *U. S. Bur. Labor Stat. Bull. no. 488*, June, 1929. Deaths from lead poisoning among painters were 50 per cent of the total deaths recorded, p. 1.

when ready mixed with oil; (2) direct handling of white lead; (3) dry rubbing or sandpapering of painted surfaces;[1] and (4) required provisions of the usual means for cleanliness. This decree in 1904 was extended to all kinds of painting with use of white lead. Finally, in 1909 a law, to take effect in 1914, was passed, prohibiting the use of white lead altogether. Other European countries studied the question of lead poisoning carefully, and enacted various regulations applicable to painters.[2]

Within recent years a decided change has come about in some aspects of the painters' trade, adding decidedly to the dangers. This innovation has been the use of the so-called spray gun for applying paint.[3] The spray method of paintting and varnishing has increased in popularity with pronounced rapidity during the past few years. The danger is obvious. A finely divided spray of lead paint, or of paint which is led-free but contains volatile thinners, such as benzol, is used and poisons the air which the sprayer must breathe.[4]

From the mechanical and production standpoints the introduction of the spray gun or airbrush has proved a most important advance in many branches of industry. In various types of factories it is the only means employed for placing permanent coatings for protective or artistic effect on manufactured articles. The three outstanding hazards in spray painting as determined by careful studies carried out under the direction of the National Safety Council and others, arise from the use of benzol, silica and lead.[5] It is gen-

[1] For use of water and mineral oils in rubbing see: *U. S. Bur. Labor Stat. Bull. no. 120,* "Hygiene of the Painters' Trade," May, 1913, p. 35.

[2] Legge and Goadby, *op. cit.,* pp. 291-293.

[3] *Final Report of the Committee on Spray Coating,* Chemical Section, National Safety Council (Chicago, Sept., 1927), p. 7.

[4] Hamilton, *op. cit.,* p. 192.

[5] *Final Report of Committee on Spray Coating,* pp. 13-14.

erally conceded that many cases of chronic benzol poisoning, plumbism and silicosis are undetected, pass for other diseases, or at least, are not reported as occupational diseases. It has been found that it would be safest to discontinue the spraying of any paints or lacquer coats containing over one per cent of lead.[1]

### PRINTING TRADES

One of the most important of the industries using metallic lead is the printing trade, with its allied branches, electrotyping, stereotyping and typefounding. The unhealthy features of the industry are due to the fact that it is an indoor occupation often carried on in hot, stagnant air; the nervous strain is considerable, especially in newspaper work and on the linotype machines; and the printer is exposed to various poisonous substances, of which lead is the most important. The amount of lead which the printer gets on his hands is small; yet though lead poisoning is rare as a cause of death among printers, it is apparently an important cause of ill health.[2]

Within recent years changes in the printing industry have had a salutary effect upon the health of the workers. The shorter hours now prevailing in printing plants are lessening industrial fatigue. Higher wages are yielding better nutrition, thereby producing a decidedly higher degree of disease resistance. Also, there is now a more general conformity to standards of personal hygiene. Still, the potential dangers of ill health, and especially liability to lead poisoning, are always present.[3]

[1] Smyth, H. F. and Smyth, H. F., Jr., "Spray Painting Hazards as Determined by the Pennsylvania and the National Safety Council Surveys," *Jour. Ind. Hyg.*, vol. x, no. 6, June, 1928, pp. 204-205.

[2] Kober and Hayhurst, *op. cit.*, p. 454.

[3] *U. S. Bur. Labor Stat. Bull. no. 427*, "Health Survey of the Printing Trades, 1922-1925," p. 110.

Lead poisoning, when it occurs in printers, is of a slow, chronic, insidious form, not easily recognized because not typical. The chief injury done by lead is probably to be found, not in the production of true plumbism, but in a lowering of resistance to other diseases, especially to certain infections. In this way is largely explained the high death rate from tuberculosis which has been noted for many years.[1]

Several state and federal inquiries have been made in the United States concerning health conditions in the printing trades. A study of the records of the International Typographical Union shows an increasing improvement in the health and longevity of the membership over a period of many years, the average age at death having increased from 41.2 years in 1900 to 58.6 years in 1928. However, tuberculosis has continued as a large cause of death among printers, diseases of the respiratory system having been the leading cause of death in 1928.[2] Wherever the death rate is high, the cause has been found in a disproportionately large number of deaths from pulmonary tuberculosis.[3]

According to Hoffman[4] in the ten years ending 1923, of 1,442 deaths from chronic lead poisoning in all industries in the population at large, only 64 were deaths of printing-plant employees, who died at an average age of 45.7 years, as compared with 49.4 years for all deaths from lead poisoning. Hoffman states that under modern working conditions fatal chronic lead poisoning in the printing trades has become a health factor of relatively small importance.[5]

[1] *U. S. Bur. Labor Stat. Bull. no. 209,* "Hygiene of the Printing Trades," 1917, p. 8.

[2] *U. S. Bur. Labor Stat. Monthly Lab. Rev.,* vol. 27, no. 5, Nov., 1928, p. 65. Also *Monthly Labor Rev.,* vol. 26, no. 4, April, 1928, pp. 72-74.

[3] *U. S. Bur. Labor Stat. Bull. no. 209,* pp. 69, 73.

[4] *U. S. Bur. Labor Stat. Bull. no. 426,* "Deaths from Lead Poisoning," 1927.

[5] *U. S. Bur. Labor Stat. Bull. no. 427,* pp. 113-115.

These views of Hoffman have been disputed by McCord,[1] who has noted that although the report of Hoffman indicates a diminishing death rate from lead poisoning, it should not be maintained that the incidence of this disease has reached negligible numbers. McCord further holds that for many reasons mortality statistics are of little value in a study of lead poisoning incidence. One reason is that rarely does a lead-exposed worker die of lead poisoning which is uncomplicated. The immediate cause of death is usually some chronic lesion to which lead absorbed over a long period may have contributed. However, the physician making out the death certificate is generally prone to place emphasis on the apparent cause of death, such as nephritis, the cardio-vascular diseases, etc., without associating lead as a primary producing factor.[2]

It appears that lead poisoning, or at least lead absorption, is still a definite hazard in the printing trades, and that much still remains to be done not only to reduce the mortality, but likewise to control the morbidity. Through a proper registration of lead workers many ailing workers could be discovered.[3]

### MISCELLANEOUS LEAD INDUSTRIES

Lead mining is carried on chiefly in southeastern Missouri, Utah, Idaho and Montana.[4] The rate of lead poisoning in

[1] *U. S. Bur. Labor Stat. Bull. no. 460*, McCord, Carey P., "A New Test for Industrial Lead Poisoning," 1928, pp. 2-3.

[2] *U. S. Bur. Labor Stat. Bull. no. 120*, p. 48.

[3] In Ohio, from the application of the occupational disease act on July 1, 1921 to January 1, 1927, 907 cases of industrial lead poisoning were reported, including 37 deaths. See: *U. S. Bur. Labor Stat. Bull. 460*, p. 3.

[4] *U. S. Bur. Mines Tech. Paper 389*, "Lead Poisoning in the Mining of Lead in Utah," 1926, states that: "Lead Poisoning contracted in the mining of lead ores is much more common than has been believed," p. 40.

a mining community depends upon the dryness of the mine, and on the proportion of carbonate and oxid ores, which are more soluble than sulphid ore, and therefore more poisonous. The deeper the mine the greater the proportion of sulphid.[1] Detailed information is not available in the United States on the subject of plumbism among lead miners.[2] Hoffman, in his study of deaths from lead poisoning, found record of 35 deaths among miners from chronic lead poisoning, and of these only three were workers in lead mines.[3] The incidence of tuberculosis among lead miners seems to be rather high in this country, and similar conditions have been found to prevail in England.[4] The danger from lead mining is generally regarded as slight, unless the ore contains lead in the form of carbonate or sulphate. It is not improbable, however, that the subtle forms of lead poisoning may occur in all lead-ore mining from the ingestion of lead dust.[5]

In the smelting of lead ore there is a distinct risk of plumbism. During the smelting process fumes are given off which are rich in lead. The greatest danger the smelter is exposed to is in cleaning out the flues of the smelting furnaces. Men enter them and remove the deposits; it is a dusty occupation, and even though the workers when thus employed wear respirators, the fume dust passes through the respirator.[6]

The processes which involve the risk of lead poisoning

[1] Hamilton, *op. cit.*, p. 124.

[2] International Labour Office, *Occupation and Health Brochure no. 128* (Geneva, 1928), p. 10.

[3] *U. S. Bur. Labor Stat. Bull. no. 426*, p. 17.

[4] Oliver, *op. cit., Dangerous Trades*, p. 285.

[5] Kober and Hayhurt, *op. cit.*, p. 459.

[6] Kober and Hayhurst, *op. cit.*, p. 414. Also, Hamilton, Alice, "Lead Poisoning in the Smelting and Refining of Lead," *U. S. Bur. Labor Stat. Bull. no. 141*, 1917.

among pottery workers are:[1] mixing the lead glaze; dipping the ware in glaze, or pouring glaze over it, or painting it with glaze; handling the ware while the glaze is still wet; removing the excess glaze from dry ware; decorating the ware by color spraying or color dusting; and cleaning or sweeping dusty floors, boards and benches.[2] In a study conducted by the United States Public Health Service, it was found that of 1,809 potters who were given a medical examination, 139 positive cases of lead poisoning were revealed, equivalent to a rate of 77 per thousand. The rate was highest among the dippers, among whom the rate of 158 per 1,000 was discovered; the glaze mixers stood next with 98 per 1,000 employed.[3] Conditions have somewhat improved since this investigation was made, however.

The manufacture of storage batteries has generally been found to be a dangerous lead trade, necessitating governmental control in Great Britain and in nearly all European countries.[4] It has been found that all operations required for the manufacture of storage batteries, expose the workmen to lead poisoning either by direct contact with lead and its compounds, or by the inhalation of lead fumes and dust.[5] In a special bulletin on the subject published by the U. S. Bureau of Labor Statistics in 1915, it was noted that of 915 workers in five large factories, 164 cases of chronic lead poisoning were discovered, equivalent to a rate of 17.9 per cent

[1] For discussion of lead poisoning among potters 200 years ago, see: Ramazzini, *op. cit.*, p. 63.

[2] Hamilton, *Industrial Poisons*, pp. 178-181.

[3] *U. S. Public Health Service Bull. no. 116*, " Lead Poisoning in Pottery Trades," 1921, p. 122.

[4] Hamilton, *op. cit.*, p. 166. See also: *Jour. Ind. Hyg.*, vol. ix, no. 8, Aug., 1927, p. 346.

[5] International Labour Office, *Occupation and Health Brochure no. 82*, p. 4.

of the number of persons employed.[1] Information derived from other sources seemingly indicates an even higher frequency of cases of lead poisoning in the manufacture of electric storage batteries, though the chronicity of the condition may not be as severe as noted in the study published in 1915.[2] Due to the study of the lead hazard in this industry, various effective measures have been taken by some of the larger companies to control the hazard, with the result that at least in some of the states lead poisoning has materially decreased in this industry in recent years.[3]

It would be quite impossible to give a full list of the occupations in which lead poisoning may occur. Lead poisoning is found in the manufacture of patent leather;[4] file cutting when done by hand;[5] polishing diamonds and other precious stones;[6] working of amber, through use of lead blocks on which the amber is cut;[7] and in many other occupations.

[1] *U. S. Bur. Labor Stat. Bull. no. 165,* "Lead Poisoning in the Manufacture of Storage Batteries," 1915, p. 23.

[2] *U. S. Bur. Labor Stat. Bull. no. 426,* p. 25. For a recent study indicating the need of periodic physical examination with particular emphasis on the study of the blood of the workers see: Greenburg, L., "A Study of Lead Poisoning in a Storage Battery Plant," *U. S. Public Health Rept.,* vol. 44, no. 28, July 12, 1929, pp. 1066-1098.

[3] Kober and Hayhurst, *op. cit.,* p. 463. See also: *U. S. Bur. Lab. Stat. Bull. no. 488,* June, 1929, indicating that while deaths from lead poisoning in storage battery plants are rare, lead absorption is extremely common, p. 3.

[4] Kober and Hayhurst, *op. cit.,* p. 677.

[5] Lawes, Edw. T. H., *The Law of Compensation for Industrial Disease* (London, 1909), p. 186. See also: Oliver, *op. cit.,* p. 343.

[6] International Labour Office, *Occupation and Health Brochure no. 21,* 1925, p. 4.

[7] *Ibid., no. 63,* 1926, p. 55.

### SUSCEPTIBILITY TO LEAD POISONING

Some lead workers appear to resist lead poisoning better than others. It may be possible that such have better eliminative capabilities. However, marked susceptibility is by far the general rule. Probably not more than one in ten workers shows any appreciable degree of natural tolerance.[1] Lead does not differ from other drugs to which persons show marked idiosyncrasies. Age and sex may, though, be regarded as predisposing factors leading to lead poisoning.[2]

As regards sex, all authorities are agreed that in women lead poisoning assumes a more severe form than in men.[3] Further, menstrual disorders have been found to be very frequent among such women workers.[4]

### LEAD POISONING AND LEAD ABSORPTION

The general consensus of opinion among qualified experts seems to be that lead poisoning continues to be the outstanding severe occupational disease in the United States. The American Public Health Association has a Committee on Lead Poisoning composed of experts in this particular field. In a report submitted in October 1928, after a study of the available evidence, the Committee similarly came to the same conclusion. The report further noted that the incidence of lead poisoning is far higher than is commonly believed or reported to state departments of health, compensation boards, etc.[5]

Much confusion now exists as to the relation between

[1] Kober and Hayhurst, *op. cit.*, p. 379.

[2] Legge and Goadby, *op. cit.*, p. 34.

[3] For diverse opinion see: Hoffman, *The Decline of Lead Poisoning,* p. 11.

[4] International Labour Office, *Brochure no. 54,* 1926, p. 4.

[5] " Lead Poisoning in the United States," Report of Committee on Lead Poisoning, *Amer. Jour. Public Health,* vol. xix, no. 6, June, 1929, pp. 631-634.

" lead poisoning " and " lead absorption ".  " Lead absorption " has come to be accepted by some as synonymous with " mild lead poisoning ", reserving the term " lead poisoning " for the profound episodes, such as encephalitis.  This interpretation has led to evasion in reporting such cases in some states requiring the reporting of " lead poisoning ", but not specifying " lead absorption ".  Hayhurst has especially pointed out this situation as prevailing in Ohio.[1]  The same difficulty has been noted in New York.[2]

In order to meet the situation thus created various efforts have been made, both to develop effective and yet easily applied tests for industrial lead poisoning, and at the same time to standardize terminology and procedures.[3] [4]  With the agreement on terminology and the acceptance of standards of diagnosis, there should be less difficulty not only in diagnosis and in the recognition of the dividing line between lead absorption and lead poisoning, but likewise a better appreciation of the actual number of workers who are mildly or otherwise affected by lead.[5]

### SUMMARY

At the present time only 11 states and the District of Columbia provide compensation and medical relief for those exhibiting signs of lead poisoning.  The states in which compensation is allowed are the following: California,[6] Con-

[1] *U. S. Bur. Labor Stat. Bull. no. 460*, p. 3.

[2] *Annual Report Industrial Commissioner*, N. Y. State Dept. of Labor, 1927, p. 191.

[3] *U. S. Bur. Labor Stat. Bull. no. 460*, pp. 32-33.

[4] Kehoe, R. A. and Thamann, F., " The Behavior of Lead in the Animal Organism," *Amer. Jour. Pub. Health*, vol. xviii, no. 5, May, 1928, pp. 555-564.

[5] Badham, Charles, *Studies in Industrial Hygiene*, no. vii, Report of the Director-General of Public Health, New South Wales, 1925, p. 69.

[6] *Stats.*, 1917, ch. 586, as amended 1919, ch. 471.

necticut,[1] Illinois,[2] Kentucky,[3] Massachusetts,[4] Minnesota,[5] New Jersey,[6] New York,[7] North Dakota,[8] Ohio[9] and Wisconsin.[10]   In addition to the District of Columbia,[11] federal civil employees are also covered by occupational disease compensation laws.[12]   With but four exceptions,[13] all of the states now provide compensation and medical relief for industrial injuries.   Although the disability from lead poisoning may be as genuine as that from non-poisonous injuries, no provision is made by the unlisted states to grant any relief.

## ANTIMONY

Antimony is a brittle, silver-white metal which is largely used as an alloy in the making of type metal in the following proportion: 60 per cent lead, 25 per cent antimony and 15 per cent tin.   Other alloys containing varying proportions of antimony are also used.   In the manufacture of these alloys and in the remelting of old and scrap metal, vapors of anti-

[1] *Gen. Stats.*, 1918, sec. 5388, as amended 1927, ch. 307, sec. 7.

[2] *Acts of 1916*, ch. 33, sec. 1, as amended 1918, ch. 176; *1922*, ch. 50; *1924*, ch. 70.

[3] *Rev. Stat. 1917*, ch. 48, sec. 154, p. 1469.

[4] *Gen. Laws 1921*, ch. 152.

[5] *Acts of 1921*, ch. 82, pt. 2, sec. 67.

[6] *Acts of 1911*, ch. 95, added 1924, ch. 124, sec. 2.

[7] *Consol. Laws*, ch. 67, added by 1914, ch. 41, as amended 1920, ch. 538; 1922, ch. 615; 1928, ch. 754; 1929, ch. 298.

[8] *Acts of 1919*, ch. 162, sec. 2, as amended 1921, ch. 142; 1925, ch. 222.

[9] *Gen. Code*, sec. 1465-68a; added 1921, p. 181, as amended 1929.

[10] *Stats. 1923*, sec. 102.

[11] 45 Stat. 600 and 44 Stat. 1424.   Under Federal Longshoremen's and Harbor Workers' Compensation Act see: 44 Stat. 1424.

[12] Acts of 1915-1916, 39 Stat. at Large, 742, sec. 40, as amended 1924, ch. 261.

[13] Arkansas, Florida, Mississippi, South Carolina.   In North Carolina Workmen's Compensation went into effect, July 1, 1929.

mony or antimony oxide are evolved.[1]   In some respects antimony resembles arsenic in its action, and in others it resembles lead.   Like arsenic it has an irritating action on the skin and mucous membrane, causing eczemas and other forms of dermatitis, and also inflammation of the mucous membrance of the mouth and nose.   It also effects the gastro-intestinal tract, with symptoms very similar to those set up by lead.   Practically all the antimony used in industry in this country contains some arsenic.   Also, since it is almost always used in conjunction with lead, some of the occupational poisoning which results from its use may easily present a mixed and not a clear clinical picture.[2]

Antimony is an important metal in only three American industries, including the printers' trade, with typefounding, stereotyping, electrotyping, monotype casting and linotype and hand composition;[3] the manufacture of grids for electric storage battery plates, especially in molding with lead-antimony alloy; and the compounding of red rubber goods.[4] The fact that the symptoms of antimony poisoning are so like that of lead poisoning makes it very hard to discover how much antimony is responsible in a case of occupational disease which occurs in a worker exposed to both metals, as in the instance of printers.   With more careful analysis of cases of industrial disease and more exact methods of determining antimony poisoning, it seems quite likely that many more cases will be uncovered than heretofore.[5]

[1] Kober and Hayhurst, *op. cit.*, p. 583.

[2] Hamilton, *op. cit.*, p. 296.   In an examination of printers in 150 shops, Hamilton could discover no clear-cut evidence of antimony poisoning, with the exception of two severe cases of eczema of the hands.

[3] Hamilton, *Jour. Ind. Hyg.*, vol. i, no. 2, June, 1919, p. 94.

[4] *U. S. Bur. Labor Stat. Bull. no. 179*, "Industrial Poisons used in the Rubber Industry," October, 1915, p. 17.

[5] *U. S. Bur. Labor Stat. Bull. no. 209*, pp. 28-29.

## ARSENIC

Arsenic poisoning in industry is generally considered first as poisoning by dust from solid compounds; and second, poisoning by the absorption of arsenic in the form of gas. Pure metallic arsenic is considered to be innocuous, while the principal poisoning compound is white arsenic, or arsenious acid.

Arsenic, in a way in which no other substance acts, has the three qualities which distinguish poisons. These are: (1) having a superficial action on the skin; (2) being absorbed by the blood, changing its constitution; and (3) having an internal, remote affect after absorption on organs and tissues, such as the heart or nervous system.[1] Among some of the industries in which arsenic poisoning has appeared are: the manufacture of arsenic products; sheepdip; paint and color works; chemical works; smelting of metals, particularly copper and lead; wall-paper manufacture; tanning and shot making. Paris green, which is aceto-arsenate of copper, is used very widely in the United States and England for spraying. This is the most important arsenical industry in the United States.[2]

In a study conducted in New York in 1917, out of 42 workers in Paris green, 24 were found to have definite symptoms of arsenic poisoning, particularly ulceration, irritation of the nasal mucosa, conjunctivitis and occasional marked anemia.[3] The manufacture of Paris green is one of the industries in which a rapid labor turnover makes for the protection of the workers against marked arsenic poisoning. A number of instances have been reported of poisoning by arsine or arseniuretted hydrogen, with a fatal outcome in ap-

[1] Legge, T. M. in Kober and Hayhurst, op. cit., Arsenic Poisoning, pp. 304-305.

[2] Hamilton, op. cit., p. 216.

[3] N. Y. State Dept. of Labor Bull. no. 83, July, 1917, pp. 11-16.

proximately 20 per cent of the cases.[1] The toxic agent in practically all the cases was produced during the interaction of metals and acids which contained arsenic. Whenever hydrogen gas is generated in a chemical reaction and arsenic is present, a portion of the arsenic is evolved as arsine.[2] In the fabrication of sheet metal products the material is frequently " pickled " by immersing the steel sheets in a bath of dilute acid in order to cleanse the metal and to remove the surface layer of rust. So long as the acid used is free from arsenic, no danger of arsine poisoning exists.[3]

### BRASS

Brass is an alloy of copper and zinc, usually containing lead in proportions varying from less than one per cent to 13 per cent. The finer varieties, which commercially are sometimes known as bronze (true bronze is an alloy of brass and tin), are usually composed of two parts of copper to one part of zinc. Brass poisoning, or brass founders' ague, is according to the present consensus of opinion, caused by the zinc in the alloy and not by the copper.[4] The essential health hazard from brass is the inhalation of the metallic fumes, an analysis of which has shown that zinc oxide predominates, with traces of iron and copper oxide also present.[5] Chronic bronchitis, emphysema and pulmonary tuberculosis have been found to be very common among brass workers. So-called

[1] Guelman, I., "Industrial Poisoning by Arseniuretted Hydrogen (Arsine)," *Jour. Ind. Hyg.*, vol. vii, no. 1, Jan., 1925, p. 6.

[2] Hamilton, *op. cit.*, p. 228.

[3] Muehlberger, C. W., Loevenhart, A. S., O'Malley, T. S., "Arsine Intoxication," *Jour. Ind. Hyg.*, vol. x, no. 5, May, 1928, pp. 137-138. Also: Jones, N. W., "Arseniuretted Hydrogen Poisoning," *Jour. Amer. Med. Assn.*, vol. 48, no. 13, Mar. 30, 1907, pp. 1099-1105.

[4] Drinker, Philip, "Certain Aspects of the Problem of Zinc Toxicity," *Jour. Ind. Hyg.*, vol. iv, no. 4, Aug., 1922, p. 177.

[5] Oliver, *op. cit.*, p. 456.

" brass chills " are to be noted among nearly all engaged in working with brass for any length of time, especially among those exposed to the fumes.[1]

### IRON AND STEEL

The chief occupational hazards encountered in the iron and steel industries may be enumerated as follows: (1) heat stroke (apoplexy), heat exhaustion, heat cramps, heat anemia and premature senility in older workers; (2) asthma, bronchitis, tuberculosis and hemorrhages while at work, due to dust, dirt, sand, gas and fume inhalations, grinding processes, etc.; (3) gas poisoning, producing chronic symptoms, such as headaches, dizziness, vomiting, anemia, palpitation, insomnia, general debility, mental dullness and occasional depressive insanity; (4) conjunctivitis; injected or blood-shot eyes due to heat, sand and dust, with cataracts among those subjected to white heated metals; (5) " sunburn " of arms, hands and face, due to exposure to heated metals; small hemorrhages under the skin of the face; (6) rheumatism and lumbago, due to great temperature variations; (7) heart disease and a number of other conditions.[2]

There are a considerable number of industries utilizing iron and steel in the manufacture of finished products, in which various additional hazards as well as occupational diseases are present. Blacksmiths and forgemen, apart from hard work and exposure to intense heat and abrupt changes in temperature, are also exposed to the inhalation of coal dust, smoke, fuel gases and fumes evolved during the tempering and case hardening with lead, potassium cyanide and oil. In the grinding and polishing processes of the metal industry in general, and in the manufacture of cutlery, tools

[1] Legge, T. M., *The Health of Brass Workers*, in Kober and Hayhurst, *op. cit.*, p. 354.

[2] Hayhurst, E. R., *Industrial Health Hazards and Occupational Diseases in Ohio*, Feb., 1915, pp. 300-301.

and steel implements, immense quantities of dust are evolved, not only from the metallic surfaces, but also from the numerous grindstones, revolving wire brushes, emery and corundum wheels, and other buffing material. File cutting, especially when done by hand, involves a severe muscular strain. In addition, while the file is being cut it is usually held upon a leaden bed to prevent slipping and to protect the surface already cut from injury. During this process considerable lead dust is produced, with resultant cases of lead poisoning.[1]

In the automobile industry, one of the most important in this country utilizing the products of iron and steel mills, there are many hazards which are not necessarily peculiar to the particular industry, but which in manufacturing the finished product utilize numerous processes which are more or less hazardous. Among these might be included iron founding, brass founding, core making, metal grinding, brazing, soldering, welding, polishing and buffing, acid dipping, steel pickling, painting and varnishing, etc. In the Ohio study of industrial hazards and occupational diseases, an analysis of the automobile industry indicated instances, among other things, of conjunctivitis, dematitis, gas poisoning, turpentine poisoning, etc.[2]

### MANGANESE

Manganese is a hard, brittle metallic element, which is very difficult of fusion. It is used as an alloy with steel or nickel, as well as in the manufacture of dry batteries; in glass works as a coloring and discoloring agent; in the cement and ceramic industries for glazing and coloring purposes; in the production of anilin colors; as a dryer in the varnish and oil industries; and for coloring soap. Manganese enters the

[1] *Ibid.*, p. 298-299.
[2] *Ibid.*, p. 59.

system in the form of dust through the respiratory organs, and also through the gastro-intestinal tract.[1]

The first American report of manganese poisoning was presented in 1913.[2] In 1919 Edsall, Wilber and Drinker published their article on manganese poisoning.[3] They found that chronic poisoning occurs as a result of inhalation and swallowing of manganese dust, the shortest period of exposure before the appearance of definite symptoms being one and one-half months, but permanent crippling required an exposure of over four months. Other cases in this country have been reported, in which the manganese poisoning occurred in the manufacture of manganese steel, during the process of which the manganese was first fused in an electric furnace.[4]

While the instances of manganese poisoning thus far discovered in the United States have been limited, nevertheless the hazard is ever-present and must be guarded against. In the factory studied by Casamajor, the attention of the management having been brought to the hazard, measures were subsequently taken to change the occupation of any worker who showed the slightest symptoms of maganese poisoning.[5]

[1] Edsall, D. L., Wilber, F. B., Drinker, C. K., " The Occurrence, Course and Prevention of Chronic Manganese Poisoning," *Jour. Ind. Hyg.*, vol. i, no. 4, Aug., 1919, pp. 183, 186.

[2] Casamajor, L., "An Unusual Form of Mineral Poisoning affecting the Nervous System," *Jour. Amer. Med. Assn.*, vol. 60, no. 9, Mar. 1, 1913, p. 646.

[3] Edsall, Wilber, Drinker, *Jour. Ind. Hyg.*, Aug., 1919, pp. 183-193.

[4] Davis, G. G. and Huey, W. B., " Chronic Manganese Poisoning," *Jour. Ind. Hyg.*, vol. iii, no. 8, Dec., 1921, pp. 231-238.

[5] Casamajor, L., *Manganese Poisoning*, in Kober and Hanson, *op. cit.*, p. 120.

### MERCURY

Mercury, both as a metal and in its manifold compounds, is one of the most poisonous of the metals used in industry. It is second only to lead in its effects upon the industrial population. Being volatile at low temperatures, it is readily vaporized; and being capable of direct absorption through the skin and mucous membranes, it is liable to produce many cases of industrial poisoning in proportion to its use.[1] The greatest demand for mercury comes from the manufacturers of explosives. Fulminate of mercury, a constituent of all percussion caps, is the most important industrial compound of the metal. The other uses are: metallic mercury for thermometers, barometers, and to a lessening extent, for vacuum pumps, especially in X-ray apparatus; for solder in dry battery manufacture; for the production of amalgam in gold and silver extraction; and for dentists' use. Mercury is also needed for the production of the acid nitrate used in the preparation of fur for felting, and for various pharmaceutical compounds.[2] There has been a reduction in the use of mercury for the making of mirrors, silver nitrate being generally substituted at the present time. It has been found that even under the best working conditions possible to achieve in mining, probably one or two cases of mercurial poisoning occur among every 100 workers.[3]

Due to the fairly large number of workers exposed, much attention has been placed in this country and in Europe on the hazards present in the felt hat industry.[4] The features

[1] Teleky, L., *Mercurial Industries*, in Kober and Hayhurst, *op. cit.*, pp. 481-482.

[2] Hamilton, *op. cit.*, pp. 246-247.

[3] International Labour Office, *Occupation and Health Brochure no. 3*, 1925, p. 14.

[4] For distribution of cases of mercurial poisoning in England see: Legge, T. M., " Mercurial Poisoning," in *Annual Report Chief Inspector Factories and Workshops* (London, 1918), pp. 70-72.

that have given to this industry its reputation as one of the worst of the "dangerous trades", are the presence of great quantities of fine fur in the air of workshops, and the use of acid nitrate of mercury as an aid in the felting of fur. The fine hair which form the fur of rabbits' skin and of the skin of muskrats, beavers, etc., are smooth, resilient and straight. Treatment with some chemical which makes them lump, twisted and rough, greatly aids in the felting process, and many chemicals have been shown to produce such an effect. Among them is the acid nitrate of mercury, now generally used in the preparation of hatters' fur.[1] Despite its highly poisonous character, mercury continues to be used for the purpose of carrotting the fur, and clinical records show a large number of cases of chronic mercurialism among workers in the felt hat industry.

Recently an intensive study was undertaken with a view of determining the prevalence of mercurialism and of other occupational diseases among those employed in various branches of the hatters' trade in this country. The examinations were made in Danbury, Conn., of a group of 100 men who were fairly well distributed among the different occupations. Of this group of workers, no less than 43 showed definite signs of mercurial poisoning. Considering that this represented a rate of 43 per cent of mercurialism in a group of hatters selected at random from the men at work in the trade, and not invalided, proves that the use of mercury to carrot fur is attended with very appreciable danger.[2,3] Analyses which were made in one of the largest hat factories

[1] Hamilton, *op. cit.*, p. 254.

[2] Wright, Wade, "A Clinical Study of Fur Cutters and Felt Hat Workers," *Jour. Ind. Hyg.*, vol. iv, no. 7, Nov., 1922, pp. 296, 304.

[3] Studies among hatters in Europe have indicated that between 25 and 60 per cent of the workers have been affected by the mercury used in the felt hat processes. See: International Labour Office, *Occupation and Health Brochure no. 5*, 1925, pp. 8-9.

in this country indicated an appreciable loss of the mercury used in the carrotting of the fur by vaporization or by treatment with hot water during the various processes in the making of felt hats.[1] The attempts to replace mercury have thus far been unsuccessful.[2]

### ZINC

Zinc is used in the manufacture of brass and bronze alloys, and especially in galvanizing iron. Zinc fumes, such as zinc oxide, cause a definite type of acute intoxication, known as zinc ague, or brassfounders' chills, zinc chills, metal shakes, etc. This intoxication is an acute malaria-like attack of chills, followed occasionally by fever and sweat, and appearing a few hours after the inhalation of zinc fumes. The intoxication occurs most frequently in founders who are exposed to the dense, whitish-green fumes rising from molten brass as it is poured into moulds. This pouring may take place four to six times a day, thus making it possible for the worker to breathe in a considerable amount of freshly formed fumes.[3]

Oliver has reported that zinc workers are subject to a mortality considerably above the average. Their mortality from respiratory diseases and pulmonary tuberculosis together was more than double the standard figure, and they died faster than the average from diseases of the circulatory system.[4] Similar conditions of a high mortality rate were

[1] Minot, A. S., "Estimations of Mercury in Hatters' Fur and in Felt," *Jour. Ind. Hyg.*, vol. iv, no. 6, Oct., 1922, pp. 253-254.

[2] Johnson, J. R., "The Preparation of Hatters' Fur: A Chemical Study of the Carrotting Process," *Jour. Ind. Hyg.*, vol. iv, no. 8, Dec., 1922, pp. 325-333.

[3] Drinker, P., "Certain Aspects of the Problem of Zinc Toxicity," *Jour. Ind. Hyg.*, vol. iv, no. 4, Aug., 1922, p. 178. See also: Drinker, P., Thomson, R. M. and Finn, J. L., "Threshold Doses of Zinc Oxide," *Jour. Ind. Hyg.*, vol. iv, no. 8, Aug., 1927, p. 331.

[4] Oliver, *op. cit.*, p. 141.

found in Chicago in a study of 1,761 foundry workers.[1]
Hoffman noted that the mortality rate from pulmonary tuberculosis among brass workers, who used zinc, was decidedly excessive at all ages under 65 years.[2]

## SUMMARY

In the preceding discussion of the hazards in the metal industries, only the outstanding causes of illness and industrial poisoning were presented. There are a considerable number of lesser metal poisons, either in the pure or in compound form, which were not touched upon. The effective control of industrial poisons will of necessity demand a constant watchfulness on the part of employers, and continuous research into causes and effects of known poisons. There are ever-increasing industrial hazards [3] and industrial managers, as well as engineers and physicians, must maintain constant guard against these industrial poisons.

## TOXIC GASES, VAPORS AND FUMES

The chemical industry is one of the key industries of this country. A large number of workers are employed in this industry, as well as in trades and occupations in which the products of chemical plants are utilized. According to the figures of the U. S. Census of Manufactures for 1925, there were 72,000 laborers and semi-skilled operatives in various chemical and allied industries; 141,000 in rubber factories; 275,000 in miscellaneous industries; about 200,000 in metal industries, such as lead and zinc, and other than iron and steel works; and 65,000 in petroleum refineries. These figures indicate to some extent that large num-

[1] *Report Illinois Commission on Occupational Disease*, Jan., 1911, pp. 67-68.

[2] *U. S. Bur. Labor Stat. Bull. no. 231*, June, 1918, p. 158.

[3] Flinn, F. B., "Some of the Newer Industrial Hazards," *Boston Medical and Surg. Jour.*, vol. 197, no. 28, Jan. 12, 1928, pp. 1309-1314.

bers of workers are daily exposed to various hazards and their health jeopardized in contact with the products and by-products of chemical industries.[1]

Unfortunately, reliable morbidity and mortality figures according to occupation are not available in this country at present on a country-wide basis. This lack of information makes it difficult to determine with any degree of exactness the amount of sickness arising out of and in the course of employment in chemical trades. However, the 1930 federal census, according to reports, will classify deaths by occupations for five or six large industrial states, and thus throw some further light on occupational hazards.

In England, Oliver found that the manufacture of chemicals, with the resultant hazardous gases, vapors and fumes, was one of the most unhealthy of the industries and one in which it is frequently difficult to protect the health of the worker. He further found that the number of deaths from respiratory diseases was very high among workers in these industries, being only exceeded by that of the cutlers and earthenware workers.[2]

In a health study of 916 workers in 22 separate chemical plants, conducted by the United States Public Service in 1916, the attempt was made to determine as far as possible the health of the workers through physical examinations. It was found that diseases of the respiratory passages were common, especially in the manufacture of acids, sulphates, chromates, etc.[3] Diseases of the digestive system were also found to be present in large numbers. In no other industry

[1] *U. S. Census of Manufactures*, 1925, pp. 746-856.

[2] Oliver, *op. cit.*, pp. 596-597.

[3] For findings in a study of hazards in New York chemical industries, see: N. Y. State Factory Investigation Commission, *2nd Report*, vol. ii, 1913, p. 607.

was the length of service as short.[1]   There was undoubtedly
a definite connection between the hazards of the industry and
the rapid labor turnover.[2]

The list of toxic gases, vapors and fumes is rather a long
one.   With the evergrowing importance of the chemical and
allied industries, and the zealous search for new compounds
for industrial and other use, the number is constantly increasing.

### ACROLEIN

Acrolein is a by-product formed when fats are rendered in
bone rendering plants; in oilcloth and linoleum factories;
varnish boiling plants; tallow rendering factories; soap factories; and in melting down old printers' type covered with
oily ink.   It is very irritating to the throat and eyes, and
causes irritation of the air passages with subsequent bronchial catarrh.   It is so irritating and injurious that it was one
of the gases tested for use in gas warfare.[3]

### FORMALDEHYDE

Formaldehyde, belonging to the same chemical group as
acrolein, is used as a disinfectant, and in combination with
phenol to make bakelite, a substance which is like hard rubber and is used extensively in the manufacture of telephone
installations and other electrical goods.   It is also used in the
rubber industry to hasten vulcanization, and as an intermediate in the manufacture of dyes.   Irritation of the respiratory mucosa is the most usual effect, resulting in many cases
of bronchitis and occasionally of broncho-pneumonia from
breathing the fumes.[4]

[1] *U. S. Public Health Bull. no. 162,* " A Health Survey of 10,000 Male
Industrial Workers," 1926, p. 106.

[2] Germany has had considerable experience with chemical hazards. See:
Kober and Hayhurst, *op. cit.,* p. 569.

[3] Hamilton, *op. cit.,* p. 432.

[4] *Ibid.,* pp. 432-433.

## AMYL ACETATE

Amyl acetate, also known as "zapone" and "banana oil", is a very useful solvent and ingredient of celluloid, shellac and of varnish and paint removers. It is also used as a lacquering agent in metallic ware and jewelry factories, as well as in oilcloth factories. During the war it was used as a "dope" for airplane wings.[1] Inhalation in the form of a vapor through the respiratory organs causes certain nervous symptoms, headaches, nausea, digestive disturbances and palpitation of the heart.[2]

## ANILIN

Industrial poisoning from anilin and substances closely allied to it have long been recognized in Germany where the manufacture of anilin dyes is a very important industry, and the possible effect on the workmen of the various substances used or evolved during such manufacture has been a subject of study for many years. There was no attempt in this country to compete with Germany in the manufacture of anilin colors or anilin itself until the war cut off the supply. The sudden introduction of these industries presented new problems to physicians and sanitarians. Making anilin from coal tar involves exposure to poisons which were new to the United States, and therefore the industrial establishments manufacturing and using anilin and its compounds gave but little thought to the protection of the workers against the dangers inherent in the industry.[3]

[1] *U. S. Bur. Labor Stat. Mo. Lab. Review*, vol. vi, no. 2, Feb., 1918, "Dope Poisoning in the Making of Airplanes," p. 48.

[2] Kober and Hayhurst, *op. cit.*, p. 644.

[3] When anilin was first introduced in the United States as a compound of rubber in the manufacture of tires, it caused a considerable number of cases of severe illness which were not readily diagnosed. See: *U. S. Bur. Labor Stat. Mo. Labor Review*, vol ii, no. 6, June, 1916, "Industrial Anilin Poisoning in the United States," p. 2.

The compounds that are used in making dyes are derived chiefly as by-products in coal-tar distillation and indirectly from benzine. Anilin enters the body through the respiratory tract, the mouth or the skin, but industrial poisoning, especially the severer forms, usually takes place through the skin. Anilin fumes also cause symptoms of poisoning, though generally severe cases do not occur through the fumes alone.[1] The toxicity of the separate products, as noted. is very different in degree. Mild cases of poisoning are generally accompanied by slight cyanosis, unsteady gait, a feeling of weakness, defective power of orientation and a terse, rapid pulse. In very severe cases there is sudden prostration, marked cyanosis, diminution of sensibility and death in a comatose condition, after antecedent convulsions.[2]

Complete statistics of cases of anilin poisoning are difficult to obtain for many reasons, among them being the fact that very often these cases are classed with those due to other substances.[3] Only incomplete figures are available for the United States, and such in states where legislation has been enacted extending the benefits of the workmen's compensation acts to certain cases of industrial poisoning.[4]

The number of anilin compounds produced in the United States has not been as numerous as in Germany. Much remains to be learned about the dangers of dye manufacture in this country. There are many points not as yet cleared up regarding methods of preventing industrial poisoning. The hazards are recognized and cases of industrial poisoning

[1] *U. S. Bur. Labor Stat. Mo. Lab. Review*, vol. viii, no. 2, Feb., 1919, "Industrial Poisoning in American Anilin Dye Manufacture," pp. 209-210.

[2] Kober and Hayhurst, *op. cit.*, p. 644.

[3] For statistics on anilin poisoning in Germany, see: Inter. Labour Office, *Occupation and Health Brochure no. 84*, 1928, p. 9.

[4] California, Connecticut, Illinois, Kentucky, Massachusetts, Minnesota, New Jersey, New York, North Dakota, Ohio, Wisconsin.

are being reported in different parts of the country, pointing to the urgent need of preventive measures.

## BENZOL

Benzene or benzol, as it is usually called, is one of the most important of the industrial poisons, but its use in industry is comparatively recent.[1] The petroleum derivatives, naphtha and benzine, were formerly used in the place of benzol. With the development of new plants for the production of benzene or benzol and its better solvent properties, it came to displace the former solvents to a marked degree.[2]

Benzol is used on a large scale in solvent extraction work, extracting oil and greases, and in certain other processes in which the hazard is largely under control. The largest commercial use, aside from gasoline for automobiles, is in the manufacture of rubber cement, rubber tires and shoes, brake linings, particularly for automobiles; artificial leather; lacquers and paint and varnish removers. All of these require the use of benzol in more or less open vessels or containers which have to be opened from time to time, thus exposing the workers to the fumes.[3]

The toxic or poisonous action of benzol takes place di-

[1] It was not until the outbreak of the World War that benzol poisoning to an alarming extent manifested itself in the United States as a result of the wholesale production of benzol and its derivatives as by-products of benzol manufactured for war purposes. See: *U. S. Public Health Repr. no. 1096*, July 23, 1926, "Benzol as an Industrial Hazard," by Greenburg, L., p. 9.

[2] *U. S. Bur. Labor Stat. Mo. Lab. Review*, vol. ix, no. 1, July, 1919, pp. 175-176.

[3] In the manufacture of benzol from coal and coal tar in the blending of motor fuels, and in the chemical industries, benzol is handled in closed containers and pipe systems. In these processes chronic poisoning is unlikely to occur and the chief hazard arises from acute poisoning due to carelessness in cleaning tanks, leaks in the apparatus and similar accidents. See: *U. S. Public Health Repr. no. 1096*, p. 57.

rectly upon the blood cells, destroying the cells which are the precursors of mature corpuscles. The symptoms caused by the inhalation of benzol fumes point to a poison with a selective action on the central nervous system, with a chain of symptoms including muscular tremors, salivation, violent twitchings, exhaustion, paralysis, narcosis, rapid pulse, disturbed respiration and low temperature. In extreme cases of poisoning there may be hallucinations, delirium, protracted unconsciousness and death during convulsions.[1]

Various reports are available relating to benzol poisoning in European countries.[2] In order to gain some idea of the extent of the benzol hazard in American industry, the Committee on Chemical and Rubber Sections of the National Safety Council, reporting on the hazard of benzol poisoning in industry, sent out a questionnaire in 1923 to a group of industrial establishments which were presumed to be being using benzol. As a result of these inquiries, the Committee came to the conclusion that the hazard of benzol poisoning in American industries is a serious one and constitutes one of the major problems of industrial hygiene.[3]

In a study made of benzol poisoning among women industrial workers in New York State the extent of disability due to the use of benzol was indicated. In six factories in different parts of the state 79 women were given careful physical examinations. Cases of chronic benzol poisoning occurred in conspicuous proportion in all the six plants. Of the 79 women exposed to benzol fumes, 25 or 32 per cent

[1] Kober and Hayhurst, op. cit., pp. 500-647.

[2] Inter. Labour Office, Occupation and Health Brochure no. 8, 1925, p. 5. Some cases were described in 1897 among women usirtg benzol as a solvent for rubber. Occasional instances of acute benzol poisoning have been reported as far back as 1862. (See: National Safety Council Report on Benzol, p. 114).

[3] Final Report on Benzol Committee Chemical and Rubber Sections, National Safety Council, May, 1926, p. 116.

showed blood changes sufficiently marked to indicate clearly that benzol poisoning was present.[1]

It has been found that even small amounts of benzol in lacquers now largely used in spray painting, may give rise to benzol concentrations in the air breathed by the sprayer well above the limit of 100 parts per million for continuous exposure, as set by the Benzol Committee of the National Safety Council.[2] The hazard is likewise present in chemical laboratories, particularly in those where tests are conducted in connection with rubber, paint, varnish and oil products to determine the solubility of compounds in benzol.[3]

The Benzol Committee of the National Safety Council reached certain conclusions regarding the use of benzol as a solvent. Among these were that in the rubber industry, in artificial leather manufacture, in sanitary can manufacture, in dry cleaning, and in the use of paints and varnishes, benzol is employed as a solvent or vehicle under conditions which, almost of necessity, permit more or less evaporation of the solvent into the atmosphere. To diminish exposure, enclosed processes should be used whenever possible.

The recommendation was also made that to detect incipient benzol poisoning at a stage when its effects can be minimized, all workers to be employed in processes where exposure to the fumes of benzol is involved, should be given a thorough medical examination before employment and reexamined, with systematic blood counts, once a month thereafter. No worker should be employed in a benzol process who shows signs of organic disease of heart, lungs or kid-

[1] N. Y. State Dept. of Labor, *Annual Report* (Albany, 1927), pp. 242-243.

[2] Smyth, H. F. and Smyth, H. F., Jr., "Spray Painting Hazards," *Jo. Ind. Hyg.*, vol. x, no. 6, June, 1928, p. 163.

[3] *U. S. Public Health Report*, vol. 43, no. 29, July 20, 1928, "Benzol Poisoning as a Possible Hazard in Chemical Laboratories," Bloomfield, J. J., pp. 1895-1897.

neys; hemorrhagic tendencies; anemia or any unusual blood picture.[1]

It is of course recognized that the ideal method of prevention would be the use of other and less hazardous solvents than benzol.    Progress along these lines is already indicated. A large can company has within the last few years announced that a substitute has been developed which will replace benzol in all its plants.[2]    A recent survey indicates that with the increased recognition of the poisonous nature of benzol and the deleterious effects upon the workers, substitutes are being developed and utilized, resulting in a marked decrease in the use of benzol in large industries.[3]

### CARBON DIOXIDE

Carbon dioxide exists as a natural and as an artificial product.    As the former, it is found in volcanic caves, mines of all kinds, at the bottom of deep wells, tunnels, sewers, etc. It is the product of decomposition and combustion of organic compounds, and a result of respiration of men and animals.[4] It is naturally increased by the process of breathing, the burning of miners' lamps, mine explosions and blasting operations.    While in itself it is not toxic, it sometimes occurs in mines in sufficient quantities to cause definite symptoms, including pressure in the head, ringing in the ears, disturb-

---

[1] *National Safety Council Report*, p. 119.

[2] N. Y. State Dept. of Labor, *The Industrial Bulletin*, vol. vi, no. 4, Jan., 1927, p. 119.

[3] Hamilton, Alice, "The Lessening Menace of Benzol Poisoning in American Industry," *Jo. Ind. Hyg.*, vol. x, no. 7, Sept., 1928, p. 231.

[4] Inter. Labour Office, *Occupation and Health Brochure no. 67*, 1926, notes that of 146 prisoners shut up during the war in India in 1756, in a room where the air only entered through a narrow opening, 123 succumbed in less than 12 hours; the prisoners shut up in vaults below the terrace of the Tuileries in June, 1848 died under analogous conditions. In these cases the toxicity of the air was in the main due to carbon dioxide.

ances in respiration, psychic excitement and sometimes convulsions. In severe cases, there may be delirium, followed by death.[1] A number of fatal cases have been reported of workers engaged in cleaning cider vats, in which the process of fermentation has generated carbon dioxide;[2] also in filling and emptying silos which are increasingly being used for the storage of grain, green corn being known to yield enormous quantities of carbon dioxide within a few hours after storage.[3]

### CARBON MONOXIDE

Carbon monoxide enters the system in the form of gas through the respiratory system. It has been held that it is the oldest industrial poison, for it must date from the discovery of the use of fire.[4] This gas is colorless, tasteless, non-irritating and when in a state of diffusion, odorless as well. It therefore gives no warning of its presence, and it has been held to be responsible for more deaths than all other gases put together.[5] Cases of acute poisoning at first result in a slow pulse, lowering of blood pressure, dull headache, ringing in the ears, nausea and fullness in the gastric region. In severe cases there is a bluish discoloration of the skin, spasmodic respiration, convulsions, asphyxia, sensory and motor disturbances, loss of consciousness, and sometimes paralysis and psychoses in the non-fatal cases.[6]

[1] Kober and Hayhurst, *op. cit.*, p. 147.

[2] N. Y. State Dept. of Labor, *Industrial Hygiene Bull.*, vol. iii, no. 10, April, 1927, p. 39.

[3] Hayhurst, E. R., "A Survey of Industrial Health Hazards and Occupational Diseases in Ohio," *Ohio State Board of Health Report*, Feb., 1915, p. 394.

[4] *U. S. Bur. Labor Stat. Mo. Labor Rev.*, vol. ix, no. 1, July, 1919, p. 173.

[5] Hamilton, *op. cit.*, p. 371.

[6] Kober and Hayhurst, *op. cit.*, p. 649.

In industry, blast furnace gas is very rich in carbon monoxide, resulting from the reduction of the oxide ores by means of glowing coke. In addition there is the carbon monoxide formed by the burning of the coke. In mining, the use of explosives in blasting, the burning of coal dust in mines and the gasolene engine exhausts, result in a considerable hazard. The danger from mine gas is increasing in the United States as the mines become deeper, for the deeper the mine, the more gas accumulates, the greater the difficulty of maintaining adequate ventilation, and the greater the possibility of miners being trapped in inaccessible places.[1] There are certain hazards in the manufacture of illuminating gas, though fatal accidents are rather infrequent, due to the precautions now taken. Testers and some other workers are regularly exposed to small quantities of carbon monoxide which may cause some minor effects.[2]

An increasingly important source of carbon monoxide poisoning is garage work, especially repairing, for the engines running idle poison the air unless a hose is attached to the exhaust to carry off the gas.[3] The almost universal use of automobiles for business and pleasure has greatly multiplied the hazard from carbon monoxide poisoning that exists in connection with the operation of these machines in garages and other closed places.[4] Studies made of garages and gar-

[1] Hamilton, *op. cit.*, p. 390.

[2] Inter. Labour Office, *Occupation and Health Brochure no. 126*, 1928, p. 14. According to the figures of the Local Sickness Office of Leipzig, the cases of sickness registered among workers at gas works are 20 per cent above the average, though fatal cases of monoxide poisoning are few.

[3] For a recent study on poisoning in garages, see: *U. S. Public Health Bull. no. 186*, " Effect of Repeated Daily Exposure of Several Hours to Small Amounts of Automobile Exhaust Gas," 1929, 58 pp.

[4] *U. S. Bur. Labor Stat. Mo. Labor Rev.*, " Hazards in a Garage," vol. viii, no. 6, June, 1919, pp. 238-240. See also: *Mo. Lab. Rev.*, vol. iv, no. 2, Feb., 1917, pp. 272-275.

age workers by the New York City Department of Health clearly indicated the hazard of carbon monoxide.[1] The New York State Department of Labor in 1920 made an inspection of 1,308 garages and auto repair shops in the state. One hundred and thirteen cases of asphyxiation were found to have occurred within two years, all but 12 of them outside of New York City. Only 36 of the garages inspected had a proper type of ventilation.[2] In another study by the same Department in 1927, it was reported that more than half of the garages studied showed a carbon monoxide concentration in excess of the 0.1 per cent regarded by authorities as the danger limit.[3]

It has been maintained that the majority of the cases of carbon monoxide poisoning remain unknown and that only the serious ones appear in reports. Considerable evidence is available on the hazards involved in a number of occupations and industries from carbon monoxide poisoning.[4] Further studies are needed relating to the elimination of the hazard; also measures for educating employers and employees.[5]

The several industrial poisons of which mention has been made are indicative of the hazards to which industrial workers are being constantly exposed. As already noted, there are many other gases, vapors and fumes than those mentioned, exposure to which results in industrial poisoning. It

[1] Harris, Louis I., "Clinical Types of Occupational Diseases," *N. Y. C. Dept. of Health Repr. no. 83*, Nov., 1919, pp. 4-6.

[2] N. Y. State Dept. of Labor, *Special Bull. no. 101*, Dec., 1920, p. 15.

[3] N. Y. State Dept. of Labor, *Annual Report*, 1927, p. 200.

[4] For some statistics on the incidence of carbon monoxide poisoning among steel workers, see: Inter. Labour Office, *Occupation and Health Brochure no. 4*, 1925, p. 5.

[5] The U. S. Bureau of Mines has prepared two films for educational purposes, one showing the dangers of carbon monoxide poisoning, and the other the construction, operation and care of an internal combustion engine. See: *New York Times*, June 20, 1928.

is presumed, however, that sufficient material has been presented to show the hazards to the workers, and to set forth the need of some measures of control and prevention.[1]

### MISCELLANEOUS CHEMICALS, FLUIDS, ETC.

An outstanding instance of an industrial hazard is that of phosphorus poisoning which may be brought about by phosphorus in the form of a vapor, through the respiratory organs; by means of food contaminated by fingers which have handled phosphorus and by action on the skin. Studies were made of phosphorus poisoning as early as 1838,[2] as well as in 1855 and 1863,[3] of workers engaged in the manufacture of matches. The necrosis of the jaw which resulted from work in the match industry led in 1852 to the development of safety matches.[4] Studies carried out by the United States Bureau of Labor and the American Association for Labor Legislation resulted in the enactment of a law in 1912 imposing a prohibitive tax on yellow phosphorus matches. Those interested in industrial hygiene came to feel that an important industrial hazard had thus been greatly minimized.[5]

A few years ago, it was discovered that cases of phosphorus poisoning were again beginning to appear in hos-

---

[1] Many cases of fume poisoning occurred during the late war in the manufacture of explosives. For a full discussion of this subject, see: Hamilton, Alice, " Industrial Poisons used in the Making of Explosives," *U. S. Bur. Labor Stat. Mo. Lab. Rev.*, vol. iv, no. 2, Feb., 1917, p. 177 *et seq.*

[2] Andrews, John B., *Phosphorous Poisoning*, in Kober and Hayhurst, *op. cit.*, p. 528.

[3] Arlidge, J. T., *The Hygiene of Diseases and Mortality of Occupations* (London, 1892), p. 456.

[4] Inter. Labour Office, *Occupations and Health Brochure no. 7*, 1925, p. 1.

[5] *U. S. Statutes*, Apr. 9, 1912, ch. 37, para. 2, 37 Stat. 81.

pitals. It was learned upon investigation that a type of yellow phosphorus fireworks was being manufactured which was causing phosphorus poisoning.[1] The United States Bureau of Labor undertook a study of the subject in 1925. There were 57 establishments manufacturing fireworks in the United States, according to the 1920 census, but only three were using yellow phosphorus and these plants were studied. In addition, some plants using yellow phosphorus as one of the ingredients of rat poison were also investigated. Among the employees in fireworks establishments studied there had been 14 definite cases of phosphorus necrosis, two of which had been fatal. The study demonstrated that there is still a real industrial hazard from phosphorus, although the number of workers exposed is small.[2]

There are many toxic chemicals and fluids which may result in poisoning. Among industries in which industrial poisons are to be found are the manufacture of sulphuric acid, nitric acid, soda, chloride of lime, chloride of phosphorus, sulphur chloride, phosgene, chloride of zinc, chloroform, carbon tetrachloride, nitrous chloride, nitroglycerin, iodine, bromine, artificial fertilizers, India rubber, artificial silk, celluloid, tar and many others. It will be noted that the manufacture of these products represents key industries without which modern industry and commerce would find it difficult to function. And yet each of those listed and others not mentioned utilize poisons in the manufacturing processes, with poisons as chief end-products and poisonous by-products and impurities.[3]

[1] Ward, E. F., "Phosphorus Necrosis in the Manufacture of Fireworks," *Jo. Ind. Hyg.*, vol. x, no. 9, Nov., 1928, p. 314.

[2] *U. S. Bur. Labor Stat. Bul. no. 405*, May, 1926, pp. 35-37.

[3] For a comprehensive tabulation of various industries in which there are mixed poisons to whose influence workmen are subjected see: Rand, Wm. H., "Composite Industrial Poisons: A Review," *U. S. Bur. Lab. Stat. Mo. Lab. Rev.*, vol. x, no. 2, Feb., 1920, p. 176 *et seq.*

The considerable number of workers engaged in the chemical industries and exposed to the poisons in greater or lesser degree must inevitably result in much industrial poisoning, many cases of which are not recognized; and, even if observed and treated, frequently are not reported. The extent of chemical poisoning of workers may be gauged by considering the report of Dr. Alice Hamilton, who noted that cases of sickness from industrial poisoning in American dye works alone have been found as a result of the use of approximately 50 different chemicals utilized in the making of dyes and dye intermediates.[1]

### GERMS

The germ diseases which might be charged up to occupations and to industry, though not in every instance, include occasional cases of hookworm disease, which may be found among agricultural laborers, excavators and others. It is found principally among coal miners, whose conditions of work are most favorable to the life of the hookworm parasite.[2] The important germ disease among workers, however, is anthrax. It is primarily a disease of animals, from whom it is contracted directly or indirectly by human beings. It is particularly frequent among cattle and sheep, but may also be transmitted to goats, horses, hogs, dogs, cats and certain kinds of game. The considerable variety of animals subject to anthrax aggravates the danger of the disease. Another serious factor is its world-wide distribution, with hardly a country being known to be entirely exempt.[3] In man anthrax is almost exclusively of occupational origin and

[1] Hamilton, *op. cit.*, pp. 517-518.

[2] Inter. Labour Office, *Occupation and Health Brochure no. 25*, 1925, pp. 4-5.

[3] *U. S. Bur. Labor Stat. Bull. no. 267*, " Anthrax as an Occupational Disease," July, 1920, p. 12.

probably no part of the country is exempt from the disease.[1] Included among those who have died from it in this country are hide and skin handlers and other tannery employees, longshoremen, wool sorters, hair workers, brush makers, paper makers, farmers, ranchmen, liverymen and veterinarians. Among non-fatal cases reported in several states and by a number of hospitals the same groups of occupations are strongly represented.

In Europe both private and public forces have combined to carry on an energetic campaign against industrial anthrax. In the United States, on the other hand, the problem has been given less consideration, although legislation for the reporting of anthrax, both as an infectious and as an occupational disease, is now fairly widespread.[2] The recommendation has been made that every large community should maintain an anthrax clinic, with a physician in charge who is fully informed in the matter of diagnosis and treatment.[3] For the workers engaged in industries and occupations in which the hazard of anthrax is present, governmental control involving not only the reporting of all new cases of anthrax, but also the thorough disinfection of all substances which might bear the anthrax spore, is of the utmost importance.[4]

[1] *U. S. Bur. Labor Stat. Mo. Lab. Rev.*, vol. iii, no. 1, July, 1916, pp. 1-2.

[2] *Bull. no. 267*, pp. 5-6. For a detailed study of the incidence of anthrax in Pennsylvania, including 123 cases in the years 1910-1921 inclusive, see: Smyth, H. F. and Bricker, E., "Analysis of 123 cases of Anthrax in the Pennsylvania Leather Industry," *Jour. Ind. Hyg.*, vol. iv, no. 2, June, 1922, pp. 53-62. This report notes that the 123 positive cases of anthrax included almost 12 per cent of the directly exposed workers. The mortality was over 21 per cent of the cases of anthrax.

[3] Schwartz, N., "Diagnosis and Treatment of Anthrax," *N. Y. Medical Jour.*, June 22, 1918, pp. 1171-1174.

[4] *Amer. Jour. Public Health*, vol. xx, no. 2, Feb., 1930, "Report of a Committee on Anthrax." Amer. Pub. Health Assn. notes an apparent increase in anthrax infection as evidenced by increasing reports; comparatively low fatality from tannery anthrax, and a high fatality from animal contact, pp. 159-160.

### MISCELLANEOUS INDUSTRIAL HAZARDS

There are a considerable number of physical and nervous disabilities which are the direct result of working conditions, including exposure to extremes of temperature,[1] excessive light, work under compressed air, skin irritations resulting in dermatitis and cancer, exposure to X-rays and radium with attendant severe burns, and many others. The listing of these few subjects may serve as an index of the kinds of additional hazards to which workers in many fields are being constantly exposed. Classes of workmen who have to endure dry heat of high temperatures almost constantly are puddlers and foundrymen, pottery bakers, glass blowers, bakers, cooks, workers with electric furnaces, etc.[2]

In electric welding of steel and other substances, the light produced is extremely intense, being equal to about 8,000 candle power, and, as the operation must be closely observed, the workmen must protect their eyes by proper shields. The light and heat evolved are sufficiently intense to cause a condition of the skin comparable to sunburn. While the hazard has been recognized, there have been many cases of various injuries to the eyes directly traceable to exposure to the excessive light.[3] Additional dangers have been noted, and these are related to the harmful effects of the light rays pro-

---

[1] The Metropolitan Life Insurance Co. made a study of the effect of occupation in the causation of certain diseases. The findings indicated that the death rate from pneumonia and other diseases of the respiratory system are much higher than the average among iron workers. The same was true of other workers who are exposed to extreme heat and sudden variations in temperature. Among iron workers, 15.9 per cent of all deaths were due to pneumonia, while among all occupied males the figure was only 7.7 per cent. See: *U. S. Bur. Lab. Stat. Mo. Lab. Rev.*, vol. xxvi, no. 6, June, 1928, pp. 51-52.

[2] For English figures on excessive mortality rates for pneumonia among iron and steel workers, see: *Registrar-General's Decennial Supplement*, pt. ii: " Occupational Mortality " (London, 1927), pp. lxx-lxxi.

[3] Kober and Hayhurst, *op. cit.*, p. 177.

duced in electric arc welding upon the tissues of the body, especially when the work is carried on in confined places, as in compartments of ships.[1]

### COMPRESSED AIR ILLNESS

Compressed air illness, sometimes known as " caisson disease "[2] or as " the bends ", threatens life and health when work is carried on under heavy atmospheric pressure, as in tunnels or in the construction of foundations for bridges and tall buildings. In excavating beneath the water level, it is necessary, in order to keep out the water, to force in air sometimes to nearly five times the common atmospheric pressure of 15 pounds to the square inch. Roughly speaking, for every five feet below the water line, two additional pounds of pressure are necessary. Those who work under such conditions, absorb through the lungs into the blood a quantity of the air which increases with the air pressure. The circulation also adapts itself to the unusual condition by establishing greater blood pressure to prevent the collapse of the blood vessels by reason of the heavy pressure from without.[3]

Most of the danger arises when the men are coming out of the abnormal environment. If decompression is too rapid, the blood vessels are dilated, partly by the increased blood pressure which has not had time to subside, and partly

[1] *U. S. Bur. Labor Stat. Mo. Lab. Rev.*, vol. viii, no. 5, May, 1919, pp. 245-246.

[2] Dr. Andrew H. Smith, who was the physician in charge of the men engaged in the building of the Brooklyn Bridge, was one of the early observers to draw attention to the hazard which he named "caisson disease". See: Smith, Andrew H., *The Effects of High Atmospheric Pressure* (Brooklyn, 1873), p. 27.

[3] Attention was first called to compressed air illness in 1841, when a compressed air caisson was used in sinking coal shafts in France and the workers became seriously ill. See: Hill, Leonard, *Caisson Sickness* (London, 1912), p. 112.

by formation in the blood stream of bubbles composed of the absorbed air which has not been able to pass off naturally through the lungs. The effects of the hasty decompression may produce deafness, blindness, prostration, paralysis, unconsciousness or death. It is the uncontrollable writhing characteristic of some of the more painful forms of the ailment which has earned for it the name of " bends ".[1]

Certain figures are available as to the incidence of this condition. During the building of the Pennsylvania East River Tunnels, in a period of 557 working days and about 1,000 men a day in compressed air at a time, there were reported to the medical department on the job a total of 3,692 cases of compressed air illness with 20 deaths.[2]

In Ontario, Canada, as elsewhere, there has been of recent date a growing realization of the hazards involved in working under compressed air, due in large part to increase in the amount of work carried on under such abnormal conditions. Ontario added " caisson disease " to the list of diseases for which compensation is to be paid. Out of 50 cases reported there were three which resulted in deaths. Since the enactment of this legislation the number of cases has been decreasing.[3]

### DERMATITIS

The cases of dermatitis, or so-called eczema, which arise in industry constitute a large percentage of the total number of cases of occupational disease which come up for compen-

[1] Keays, F. L., " Compressed Air Illness," *Amer. Labor Legis. Rev.*, vol. ii, 1912, p. 200. See also: *Amer. Labor Legis. Rev.*, vol. iv, 1914, pp. 547-548.

[2] Erdman, Seward, in commenting on these figures noted: " I feel certain that if every trivial case of ' bends ' had been recorded the percentage would have been doubled; indeed, I doubt that even 10 per cent escaped unscathed at all times." In Kober and Hayhurst, *op. cit.*, p. 791.

[3] Dept. of Health, Ontario, Canada, *Annual Report*, 1926, p. 21.

sation every year in some states. These eruptions may vary from a very small area of redness, to very extensive destruction of skin tissues over large areas of the body. The eruption usually begins on the hands, since they are the first to come in contact with irritant substances. There are a great variety of substances which may cause dermatitis. In addition, certain workers possess peculiar susceptibilities to some substances, and readily develop marked cases of occupational dermatitis.[1]

Certain lubricants and cutting compounds are used which are necessary for many operations where the cutting edge of a machine tool is applied to the surface of metals. These compounds vary in composition, for the most part being a combination of animal oils, mineral oils and fatty acids, mixed with water to form an emulsion. The dermatitis which occurs among those working with cutting compounds consists in a series of boils or multiple abscesses. It is believed as a result of several investigations, that infection follows lessened resistance of the skin to the irritant action of the oil, coupled with a lack of personal cleanliness.[2]

Instances of occupational skin diseases have been noted in the leather trade,[3] among dyers, anilin workers, dressmakers, photographers and others.[4] The New York State Department of Labor made a study of dermatitis cases and the causative agents reported. Out of a total of 390 cases, soaps

[1] Mayers, May R., "Skin Eruptions in Industry," *N. Y. State Industrial Hygiene Bull.*, vol. v, no. 2, Aug., 1928, p. 5.

[2] During the World War there were many instances both in this country and in England of skin conditions arising among those exposed to cutting oils used in munition and other factories. See: Yates, A. B., "Cutting Compounds as a Cause for Dermatitis and Wound Infection," *U. S. Bur. Labor Stat. Mo. Lab. Rev.*, vol. viii, no. 1, Jan., 1919, pp. 273-277.

[3] *U. S. Bur. Labor Stat. Mo. Lab. Rev.*, vol. x, no. 4, April, 1920, p. 84.

[4] Kober and Hayhurst, *op. cit.*, p. 914.

and cleansing powders were responsible for 120, dyes 47, lime (cement) 45, lead 19, ink (carbon paper) 18, oils and grease 17, chromic acid 12, methyl alcohol 11, etc., etc. The occupations were widely distributed.[1] The widespread nature of these industrial disabilities will be further recognized as the compensation laws of the various states become more liberal and make provision for compensation as well as careful reporting of occupational skin diseases.

### CANCER

Workers coming in contact with tar and pitch suffer from an inflammatory affection of the skin, so-called tar eczema, which occasionally takes on a cancerous nature. The occurrence of skin affections and cancer in the petroleum industry has been noted by several observers, especially among those employed on the unpurified mineral oils.[2] Dr. Alice Hamilton found that it is frequently difficult to gain a clear idea as to just what substances play the most important part in the skin lesions and cancers found among petroleum and tar workers. It is not always possible to determine whether the lesions are due to chemical irritants or are caused by mechanical agents, such as the plugging of sebaceous ducts or injury by sharp bits of metal and subsequent infection.[3]

Mule-spinners' cancer has been known in England for some time, though it is only of recent date that it has been recognized as a separate disease entity. The causative factors are still a matter of dispute among physicians and scientists who have been attempting to solve the connection of this type of malignant growth with the occupation, though it

[1] N. Y. State Dept. of Labor, *Industrial Hygiene Bull.*, vol. viii, no. 2, Nov., 1928, pp. 454-455.

[2] Hoffman, F. L., *The Mortality from Cancer Throughout the World* (Newark, 1915), pp. 58-61.

[3] Hamilton, *op. cit.*, p. 411.

seems to be the opinion that the tissues are affected by the petroleum oils which are used in lubricating the machines.[1] As soon as American mortality statistics are reported according to occupation, it may be possible to obtain a fairly correct estimate of this industrial hazard in the United States.[2] From present information, it seems that the apparently small incidence of cancer among American mill workers results from the use of a highly refined spindle oil which is harmless to the worker.[3]

### RADIUM HAZARD

Occupational injuries due to radium have apparently been well known to physicists working with radioactive substances, and they have generally taken necessary precautions to protect themselves.[4] The effect of the use of radioactive substances on the health of workers employed in painting the dials of watches and clocks with luminous paint has been the subject of investigation for some time, especially since the first death from this cause was reported in 1924.[5]

The subject was recently brought to the fore again through the suit brought against the United States Radium Corporation, Orange, N. J., by five women who developed

[1] Hoffman, F. L., "Mule Spinners' Cancer," *U. S. Bur. Labor Stat. Mo. Lab. Rev.*, vol. xxvii, no. 3, Sept., 1928, pp. 27, 42.

[2] For statistics on the incidence in England and Wales, see: Southam, A. H., "Occupational Cancer of Mule Spinners," *British Medical Jour.*, vol. ii (London, Sept. 8, 1928), p. 437.

[3] Heller, Imre, "Occupational Cancers," *Jo. Ind. Hyg.*, vol. xii, no. 5, May, 1930.

[4] At the Radium Institute in London the plan was put into effect some years ago of rotating monthly the nurses handling the radium applicators, in order to reduce the recognized hazard. See: Ordway, Thomas, "Occupational Injuries due to Radium," *Jour. Amer. Med. Assn.*, vol. lxvi, no. 1, Jan. 1, 1916, pp. 1-3.

[5] *U. S. Bur. Labor Stat. Mo. Lab. Rev.*, vol. xxiv, no. 11, Nov., 1925, pp. 181-187.

more or less serious conditions as a result of their work with the radioactive paint. Fourteen girls who had been employed at various times since 1917 in the plant at Orange, N. J. have died as a result of the absorption of radium through the practice of pointing their brushes in their mouths.[1] Other deaths have been reported in plants elsewhere, particularly in one at Waterbury, Conn.[2] In these cases radium necrosis attacked the jaws and fingers. In addition, the radioactive elements have a profound effect on the blood-forming centers of the human system, causing an anemia of a pernicious type. The radium which is absorbed into the system is deposited in the bones, spleen and liver.[3] The seriousness of the situation led to the calling of a conference by the United States Public Health Service and it is hoped that the necessary measures will be taken to control this occupation hazard.[4] Even with further control than is at present being exercised, it is evident that more ready means must be available to the afflicted worker than a long court suit in order to recover damages resulting from exposure to an extremely hazardous industrial poison.[5]

[1] It has recently been reported that no case of radium poisoning has come to light in Europe or elsewhere, but only in the United States, where the habit of pointing the brush with the lips developed. *The Lancet*, vol. ccxvi, no. 5511 (London, April 13, 1929), p. 786.

[2] For a full discussion of the subject see: *U. S. Bur. Labor Stat. Mo. Lab. Rev.*, vol. xxviii, no. 6, June, 1922, " Radium Poisoning," pp. 20-95. See also: Martland, H. S., *Jour. Amer. Med. Assn.*, Feb. 9, 1929, pp. 466-473.

[3] *U. S. Bur. Lab. Stat. Mo. Lab. Rev.*, vol. xxvii, no. 7, July, 1928, pp. 42-43.

[4] *New York Times*, Dec. 21, 1928.

[5] *Amer. Labor Legis. Rev.*, " New Jersey Radium Cases," vol. 18, no. 4, Dec., 1928, p. 388.

### CONCLUSION

Discussion of the various industrial hazards in the preceding pages indicated that modern industry is responsible for a considerable proportion of the workers' physical ills. How far other conditions are responsible it is difficult to state, though the definite information already available leaves no ground for doubt as to the contribution of industry to the high toll of illness and death among the working population. Dublin has noted that the degenerative diseases, such as cerebral hemorrhage, Bright's disease and organic heart disease, show strikingly the effects of industrial exposure. The death rates are two and three times as high as in non-industrial groups during the active working years of life. These high rates probably reflect the results of long-continued strenuous labor, of heat, of marked changes in temperature, and in some instances, of specific occupational poisoning. These conditions in their early stages are very difficult to diagnose and later lose the indications of their occupational origin. The true picture, according to Dublin, is probably worse than would appear from the records of mortality.[1]

It would be extremely valuable to be able to measure the economic problem of industrial disease in the United States, but unfortunately American information on the subject is largely lacking. Even in Europe satisfactory data for industrial morbidity and mortality can be found only in a few countries. A beginning has been made in this country in the gathering of exact information on the subject of occupational diseases. The information available, however, can only be found in such states as provide compensation for occupational diseases, and few, as will be noted later, make complete provisions for compensating all occupational diseases. The Massachusetts Department of Industrial Acci-

[1] Dublin, Louis I., *Health and Wealth* (New York, 1928), pp. 277-278.

dents in the annual report for the year ending June 30, 1927, notes that there were 64,167 claims for compensation, of which only 810 were due to occupational diseases, being somewhat over one per cent of the total number of claims.[1] These 810 disease claims included 8 of anthrax; 104 dermatitis; 16 eczema; 61 lead poisoning; 140 poison ivy; 12 tuberculosis, including stone cutters' consumption; 40 gas poisoning; 429 all others.

From the fact that only one per cent of the claims involved occupational diseases, it might be presumed that these diseases form only a small part of industrial accident and sickness problems.  However, students of the subject in this country and in Europe believe that industrial illness arising out of occupational diseases and industrial poisons is, in its economic effects, a much more destructive factor than industrial accidents.[2]  To meet the large problem of industrial illness, and, as far as possible, to prevent the development of new cases, various measures have been taken which will be discussed in the following chapter.

[1] Massachusetts Dept. of Industrial Accidents, *Annual Report Year Ending June 30, 1927*, p. 80.

[2] Rubinow, I. M., *Social Insurance* (New York, 1913), p. 207.

# CHAPTER IV

## REGULATION AND PREVENTION OF OCCUPATIONAL DISEASES

### INTRODUCTION

IN the historical summary of occupational diseases mention was made of certain diseases which had been recognized and were known to the ancients as well as to those of a later period. The development of the factory system in the latter part of the 18th century brought new industrial evils in its wake. In England the new system of production had been upwards of half a century in operation before public attention was directed to it in any considerable degree. During that time it had spread over a wide area. Towards the end of this period, the great inventions for the treatment of wool and cotton fibre had been coming into use, and in 1787 were likewise applied to flax. Almost immediately complaints began to be heard of abuses connected with factories. This led to the appointment of a committee of investigation at Manchester in 1795, which reported the existence of a number of evils seriously affecting the health of the workers, and especially of the children employed in the mills. A series of recommendations were drafted, the final recommendation being that an appeal be made to Parliament for the enactment of legislation to control such establishments.[1]

The year 1802 was marked by the passing of the first of the long series of statutes regulating the hours and condi-

[1] Taylor, R. W. C., *The Factory System and the Factory Acts* (London, 1894), p. 34.

tions of labor, commonly known as the Factory Acts. These Acts were framed with the definite and avowed object of protecting the health of the younger and weaker workers from injury by overwork or unwholesome conditions. It is this, their motive and purpose, rather than the actual matter of the regulations, that forms their distinguishing characteristic and marks them out as a novelty and new departure from previous legislation.[1]   This first factory act was known as the Health and Morals Apprentices' Act 1802 " (42 George III, Ch. 73).   The statute was particularly important as being the first definitely in restraint of the modern factory system and in general opposition to the *laissez-faire* policy in industry.   It applied to cotton and woolen mills, and its regulation for the most part applied only to apprentices, and to factories in which " twenty or more persons were employed; " but the clause relating to restriction of hours, as well as those for education, were expressly applied to apprentices only.   The whitewashing and ventilation clauses applied to all factories.[2]

The early factories were constructed in districts where water power was abundant and often in remote localities, where public control was rarely exercised.   However, the introduction of steam power transferred many factories to populous places, and resulted in a change in the workers, as children employed in such areas did not have to be apprenticed, and worked there accordingly without participating in the benefits of the First Factory Act.   Sir Robert Peel brought this matter under the notice of the House of Commons on June 6, 1815.   Up to this stage factory legislation had excited but little general interest, and absolutely no alarm.   Nothing had heretofore been said about the necessity

[1] Hutchins, B. L. and Harrison, A., *A History of Factory Legislation* (London, 1903), p. 1.

[2] *Ibid.*, p. 17.

of any public inquiry into this matter, a device destined to play a great part thereafter.[1] Further legislation was enacted in 1833, when four factory inspectors were provided for, appointed and paid by the central government. They were granted full powers to enter any textile mill or factory at any time, and to make inquiries as to conditions under which children and young persons were employed.[2]

Regarding the matters of health and safety, the earlier Factory Acts contain only simple provisions for whitewashing, ventilation, etc. The dangers of machinery were not taken cognizance of by law until 1844, and then chiefly owing to the representations of the inspectors. The industries most nearly connected with textiles were the first to be taken under consideration. Acts were passed dealing with print works (1845), bleaching and dying (1860) and lace works (1861). In 1863 Parliament took a further step and passed an act imposing certain regulations on bakeries; the Act of 1864 included earthenware, lucifer matches, percussion caps, cartridge making and paper staining.[3]

This act was specifically passed to deal with certain trades supposed to be especially unhealthy, but no specific precautions were enacted beyond the very general provision that " every factory to which this Act applies " should be kept in a cleanly state, and be ventilated in such a manner as to render " harmless so far as is practicable " any gases, dust or other impurity generated in the process that might be injurious to health.[4] Further legislation was enacted in 1867, 1878, 1883, 1891, 1901 and in subsequent years. The purpose of this legislation was to exclude young children from

[1] Taylor, *op. cit.*, p. 57.

[2] Webb, Mrs. Sidney, *The Case for the Factory Acts* (London, 1902), p. 115.

[3] *Ibid.*, pp. 83-84.

[4] Hutchins and Harrison, *op. cit.*, p. 20.

factories, to control the hours of labor for women and children, and to provide machinery for the inspection, reporting and control of industrial hazards and occupational diseases.

This brief summary of the English experience is presented as throwing some light upon early attempts at regulating working conditions, and protecting the health of the worker.

### REGULATION IN THE UNITED STATES

The method of regulation for the prevention of occupational diseases, as in other social problems, is based on the principle of toleration within certain limits. The majority of the people may believe that certain harmful conditions are so necessary a part of our industrial life that their prohibition is at present undesirable or at least impracticable. As a result, the method of regulation is far more widely used than the method of prohibition. The adoption of this method for the control of occupational diseases leads to limitations in three different directions: (1) upon the methods of handling the hazardous materials; (2) upon the period of exposure; and (3) upon the persons exposed. These three types of limitation are seldom used singly, but are usually combined in the same law.[1] In the United States, the several states, as well as the federal government, have enacted legislation dealing with various aspects of the problem of regulating working conditions which might result in the development of occupational diseases. These regulations have been enacted into laws under various headings. Some of these will be indicated and mention will be made of typical regulations in various states, as an index of the attempts to deal with the problem.

Among the subjects dealt with in various states, and in some instances, by all the states, are the following: (1) inspection and regulation of factories and workshops; (2) re-

[1] Andrews, John B. in Kober and Hayhurst, *Industrial Health*, p. 1005.

ports and investigation of occupational diseases; (3) mine regulations; (4) ventilation; (5) air space in workrooms; (6) employment of children and minors in dangerous occupations; (7) employment of women in dangerous occupations; (8) preventive regulations of various kinds; (9) work in foundries; (10) periodic medical examination of lead workers; (11) control of work in bakeries; (12) regulation of work in cigar factories; (13) regulation of work in compressed air; (14) use of infected or soiled wiping rags in print shops; (15) precautions when polishing wheels are employed.

### INSPECTION AND REGULATION OF FACTORIES AND WORKSHOPS

All the states have enacted laws dealing with the inspection and regulation of factories and workshops. As might be presumed, those states in which industrial establishments are widely diversified and contain many hazards, have more elaborate regulations than others in which such is not the case. In order to simplify the discussion of factory regulations as a whole, several provisions will be discussed under separate headings. The law of California, for instance, provides that: " Every factory, workshop, mercantile or other establishment, in which five or more persons are employed, shall be kept in a cleanly state. . . ." [1] This is a general provision, which is found in many states. Further sections in the California law deal with ventilation; prohibition of use of cellars as workplaces; installation of fans, blowers, etc., " where a work or process is carried on by which dust, filaments, or injurious gases are produced or generated that are liable to be inhaled by persons employed therein ".[2] There is a penalty provided for violations, with responsibil-

---

[1] *General Laws*, California. Sims' Deerings' Codes 1906. Act no. 1098, sec. 4 as amended 1909, ch. 52.

[2] *General Laws*, Act no. 1098, sec. 4, as amended 1909, ch. 52.

ity for inspection and regulation placed upon the commissioner of the Bureau of Labor Statistics.

### REPORTS AND INVESTIGATION OF OCCUPATIONAL DISEASES

Nearly all the states have regulations as to the reporting and investigation of occupational diseases. Usually the burden of reporting the cases of occupational diseases is placed upon all physicians to whose attention such cases may come. In Connecticut, the law, in part, reads as follows:

Every physician having knowledge of any person whom he believes to be suffering from poisoning from lead, phosphorus, arsenic, brass, wood-alcohol, mercury, or their compounds, or from anthrax, or from compressed-air illness, or any other disease, contracted as a result of the nature of the employment of such person, shall, within forty-eight hours, mail to the commissioner of labor and factory inspection a report stating the name, address and occupation of such patient, the name, address and business of his employer, the nature of the disease, and such other information as may reasonably be required by said commissioner. The commissioner shall prepare and furnish to the physicians of this state suitable blanks for the reports herein required. No report made pursuant to the provisions of this section shall be evidence of the facts therein stated in any action at law against any employer of such diseased person. Any physician who shall neglect or refuse to send any report herein required, or who shall fail to send the same within the time specified herein, shall be liable to the state for a penalty of not more than ten dollars, recoverable by civil action in the name of the state by the said commissioner. For each such report the physician making the same shall receive a fee of fifty cents to be paid by the State Department of Health as a part of its office expenses.[1]

An almost similar law is found in Maryland,[2] Maine,[3]

[1] Sec. 2416, as amended 1923, ch. 93.

[2] *Public General Laws*, Code of 1911, sec. 5g (added 1912, ch. 165).

[3] *Revised Statutes 1916*, ch. 19, sec. 19.

Michigan,[1] Minnesota,[2] New Hampshire,[3] New York,[4] Rhode Island[5] and Wisconsin.[6] The intent and purpose of the law is undoubtedly a good one, for, if the provisions of the law were carried out, all physicians in states where occupational diseases were apt to occur, would report such cases to the proper authorities, and they in turn might take the necessary steps to control the hazards.[7] In practice, things are not so simple. In the first place, physicians as a group are not trained to recognize the various occupational diseases, and do not always have available the necessary laboratory facilities to study cases where there is a questionable diagnosis. Secondly, the diagnosis of certain of the occupational diseases listed is frequently very difficult, and only specialists with considerable training in the treatment of some of the diseases, are competent to make a definite diagnosis, and therefore to send reliable data to the state officials. Practically, the control of occupational diseases has not been much advanced by the above-mentioned provision of the law, though widespread information and special instructions among physicians might result in early discovery of many cases.

### MINE REGULATIONS

All the states in which mining operations are carried on have laws relating to the control of employment in mines. These laws generally declare that employment in under-

[1] *Compiled Laws 1915*, sec. 5166.

[2] *General Statutes 1913*, sec. 3899.

[3] *Acts of 1913*, ch. 118, sec. 1.

[4] *Consolidated Laws 1909*, art. vii, sec. 206.

[5] *General Laws 1923*, ch. 163, sec. 23.

[6] *Statutes 1923*, secs. 69, 49.

[7] N. Y. State Dept. of Labor, *Indus. Hyg. Bull.*, vol. v, no. 11, May, 1929, states: "During the past four years less than ten physicians have reported diseases, and not more than five hospitals have notified the Dept. of Labor," p. 42.

ground coal mines, underground lode mines, and all other underground mines or workings of any kind, or in smelters, reduction works, etc., is injurious to health and dangerous to life and limb. As most of the hazard in mining lies in accidental injuries, considerable stress is naturally placed upon the proper protective construction of the mines, adequate experience of miners and the examination of their competence, various provisions for safety, duties of mine inspectors, construction and maintenance of wash houses, etc.[1] In addition, hours of employment are in most instances limited to eight per day,[2] women and children are prohibited to be employed,[3] and certain other measures have been enacted with the idea of controlling as far as possible the development of occupational diseases. Among these may be mentioned the requirement for adequate ventilation. For instance, the Indiana law stipulates that:

The operator of every mine shall provide and maintain, hereafter for every such mine, a sufficient amount of ventilation affording not less than 100 cubic feet of air per minute for each and every person employed therein, and 300 cubic feet per minute for each mule, horse, or other animal, in said mine, measured at the intake of the split or subdivision of the air, and as much more as the circumstances require. It shall be forced and circulated around main entries, cross entries, and working faces throughout the mine, so that all open places shall be free from standing gas of whatsoever kind, to such an extent that the entire mine shall be in a fit state at all times for the men working therein, and will render harmless all noxious or dangerous gases generated therein.[4]

---

[1] See: *Indiana Acts 1923*, ch. 42, for comprehensive mine regulations; also *Colorado Compiled Laws 1921*, sec. 4172.

[2] *Alaska Acts of 1917*, ch. iv, sec. 2; *California Acts of 1913*, ch. 186, sec. 1; *Colorado Compiled Laws 1921*, sec. 4173.

[3] *Alabama Code of 1923*, sec. 1724; *Colorado Compiled Laws 1921*, sec. 3482; *N. Y. Consol. Laws 1909*, art. iv (as amended 1921, ch. 642), para. 7, etc.

[4] *Acts of 1923*, ch. 42, sec. 10.

A further attempt to control working conditions which may be directly or indirectly the cause of occupational diseases is the Nevada law which stipulates that where rock drilling by machinery creates dust, a water jet or spray or similar device must be used to prevent the escape of the dust. Another requirement imposes responsibility upon the worker to utilize such appliances when the drills are equipped with water sprays.[1] These, in brief, are the important mining regulations thus far enacted to protect the health and lives of the workers, though not always with a comprehensive realization of the actual occupational diseases to which the miners are exposed.

### AIR SPACE IN WORK ROOMS

A number of states make definite requirements as to the amount of air space to be provided per worker in factories, mercantile establishments, mills, etc. The matter of air space is integrally tied up with the question of ventilation. The following is the wording of the Illinois law relating to this matter:

In every room or apartment of any factory, mercantile establishment, mill, or workshop, where one or more persons are employed, at least five hundred (500) cubic feet of air space shall be provided for each and every person employed therein, and fresh air, to the amount specified in this act, shall be supplied in such a manner as not to create injurious drafts, nor cause the temperature of any such room or apartment to fall materially below the average temperature maintained, Provided, Where lights are used which do not consume oxygen, 250 cubic feet of air space shall be deemed sufficient.[2]

---

[1] *Acts of 1913*, ch. 125, secs. 1 and 2. There are some doubts at present as to the efficacy of this method of controlling the fine particles of dust which actually do the damage, causing silicosis or other lung conditions.

[2] *Revised Stat. 1917*, ch. 48, sec. 99. See also *New Jersey Compiled Stat. 1910*, sec. 35; *Penna. Stat. 1920*, sec. 13593.

In the requirement for adequate air space there is a realization of the occupational hazard inherent in the lack of sufficient clean air in workplaces accommodating a number of workers. A further attempt to deal effectively with the problem of polluted air in workrooms is indicated by the requirements of the New Jersey statute dealing with the ventilation required in definitely hazardous workrooms. The statute reads, in part, as follows:

If excessive heat be created or if steam, gases, vapors, or dust or other impurities that may be injurious to health be generated in the course of the manufacturing process carried on therein, the rooms shall be ventilated in such a manner as to render them harmless, so far as is practicable. If glazing or polishing on a wheel or any process is carried on by which dust or any gas, vapors or other impurity is generated in such manner as to be inhaled by the employees to an injurious extent, and it appears that such inhalation could be to a great extent prevented by a fan or other mechanical means, the commissioner of labor may order the owner, agent or lessee of such place to provide a fan or other mechanical means of a proper construction for preventing such inhalation within twenty days after the service upon him of such order in writing.[1]

### CHILDREN AND MINORS IN DANGEROUS OCCUPATIONS

Mention has already been made of the prohibition of employment of children and minors in the hazardous occupation of mining. There are a number of other occupations, hazardous in nature, to which similar prohibitions apply. Wisconsin has a most comprehensive law regulating the employment of young persons in occupations in themselves hazardous, or which might result in the development of occupa-

---

[1] *Compiled Stat. 1910,* sec. 35 (as amended 1912, ch. v). See also, for similar provisions: *Conn. General Stat. 1918,* sec. 2350; *Del. Acts 1917,* ch. 231, sec. 6; *Illinois Revised Stat. 1917,* ch. 48, secs. 43-46; *Mass. General Laws 1921,* ch. 149, secs. 117-120; *N. Y. Consol. Laws 1909,* art. xii, sec. 299.

tional diseases. The following are among the employments deemed to be dangerous or prejudicial to the life, health or safety of minors under eighteen years of age, and in which they are prohibited from being engaged; in or about blast furnaces; operating or using any emery, rouge, corundum, stone, carborundum, and abrasive or emery polishing or buffing wheel, where articles of the baser materials are manufactured; in or about establishments where nitro-glycerine, dynamite, guncotton, gunpowder or other high or dangerous explosives are manufactured or compounded; in dipping, drying or packing matches; in or about mines or quarries.

Minors under sixteen years of age are prohibited in occupations involving burnishing machines; in any tannery or leather factory; occupations causing dust in injurious quantities; emery or polishing wheels for polishing metals; manufacture of paints, colors or white lead; manufacture of any composition in which dangerous or poisonous acids are used; manufacture or preparation of compositions of dangerous or poisonous dyes; manufacture or preparation of compositions with dangerous or poisonous gases; manufacture or preparation of compositions of lye in which the quantity thereof is injurious to health; washing, grinding or mixing mill or calendar rolls in rubber manufacturing; in any tobacco warehouse, cigar or other factory where tobacco is manufactured or prepared; woodshaper, planer, sandpaper, wood polishing or woodturning machine; wool, cotton or hair upholstering.[1] It is interesting to note the number of occupations which are considered hazardous to the health and lives of young people. Quite naturally, the same hazards may be presumed to be dangerous to the well-being of adults, though perhaps, not always to the same degree.

---

[1] *Wisconsin Stat. 1923*, sec. 103.05, para. 3. Nearly all the states have similar provisions for the protection of minors. See especially *Del. Revised Code 1914*, ch. 3145, sec. 45 (as amended 1923, ch. 202).

### WOMEN IN DANGEROUS OCCUPATIONS

In a number of states the general provisions for the protection of the health of children apply similarly to women through the prohibition of the employment of both groups in certain occupations through mandatory regulation of the industrial diseases.    In other instances, women alone are specifically prohibited from being employed in occupations known to result in industrial diseases.    In Louisiana the law provides that: "In all establishments wherein children, young persons, or women are employed where any process is carried on by which dust, or smoke or lint is generated," fans or other dust, or smoke or lint-removing or consuming contrivances must be so placed as to prevent the inhalation of such dust or smoke or lint by the employees.[1]    New York law stipulates that: " No male under eighteen years of age, nor any female shall be employed in operating or using any emery, tripoli, rouge, corundum, stone, carborundum, or any abrasive or emery polishing or buffing wheel, where articles of the baser metals or iridium are manufactured." [2]

Massachusetts,[3] Minnesota[4] and New York,[5] prohibit the employment of women in core rooms, in which cores are placed into ovens or taken out therefrom.    The New York provision of the law reads as follows:

No female shall be employed in a foundry at or in connection with the making of cores where an oven in which the cores are baked is in operation in the same room or space in which the cores are made.    A partition separating the oven from the space where the cores are made shall not be sufficient, unless the partition extends from floor to ceiling and is so constructed

---

[1] *Louisiana Acts 1912*, Act no. 301, sec. 19.

[2] *N. Y. Consol. Laws 1909*, art. iv, sec. 146 (as amended 1921,, ch. 642).

[3] *Mass. General Laws 1921*, ch. 149, sec. 54.

[4] *Minn. Acts of 1919*, sec. 20.

[5] *N. Y. Consol. Laws 1909*, art. iv, sec. 147.

and theopenings therein so protected that gases and fumes from the core oven will not enter the space in which women are employed.

Several states recognize the hazard of work in foundries.[1] The law of Minnesota on this subject might be considered as a model for legislation of this type. Among the provisions of this law are the following:[2]

Where smoke, steam, gases, or dust arising from any of the operations of the foundry are dangerous to the health or eyes, and where a natural circulation of air does not carry off the greater part of such smoke, steam, gases or dust, there shall be installed and operated adequate mechanical means of ventilation.

The cleaning and chipping of castings shall be done in cleaning rooms, except that castings may, when necessary, be chipped or cleansed in the molding room or where cast, provided sufficient protection is furnished by the use of a curtain or screen, or some other means equally good, to protect employees therein. This section shall not apply if mechanical appliances are used for cleaning castings and the dust and particles arising therefrom are effectively removed.

Where tumbler mills are used, exhaust systems shall be installed to effectively carry off the dust arising from the cleaning of castings, except where the mill is operated outside the foundry. This section shall not prohibit the use of a water barrel for cleaning castings. Sand blast operations shall be carried on in the open air or in a separate room and used solely for that purpose. The milling of cupola cinders, when done inside the foundry, shall be carried on by an exhaust mill or water mill.

No cores shall be blown out of castings by compressed air

[1] *Indiana Act of 1919*, ch. 39, sec. 1, requires employer to supply suitable gas masks in employments exposing workers to dangerous, noxious or deleterious gases.

[2] *Minn. Acts of 1919*, ch. 84, secs. 6-10.

unless such work is done outside the foundry or in a special or dust proof inclosure. Employees engaged in cleaning castings by compressed air or sand blast shall wear eye guards and helmets, to be furnished by the employer.

When fumes, gases, and smoke are emitted from drying ovens in such quantities as to be detrimental to the health or eyes of the employees, hoods and pipes or other adequate means of ventilation shall be provided.

### REGULATION OF LEAD POISONING

Lead poisoning as has been heretofore mentioned, is one of the important causes of industrial disability. Several states have taken cognizance of this, and have enacted regulatory provisions aiming at the control of this widespread occupational disease hazard. Missouri among other states, includes the use of lead, in any manufacturing process, as especially dangerous to the health of the workers, and requires employers to adopt and provide approved and effective devices for the prevention of lead poisoning and other occupational diseases.[1] New York requires, as a precautionary measure, that factories in which lead, arsenic or other poisonous substances, or injurious or noxious fumes, dust or gases are present as an incident or result of the occupation, shall furnish hot water, soap and individual towels for the use of the workers.[2] New Mexico provides that any worker disabled or rendered unfitted for labor by reason of lead poisoning acquired while employed in smelting works, shall be provided by his employer with all proper medical attendance, medicine and sustenance during such disability.[3] While this provision of the law partakes of compensation payment for an occupational disease, it likewise has definite regulatory value.

[1] *Rev. Stat. 1919*, secs. 6817-6818.
[2] *Consol. Laws 1909*, art. xi, sec. 293, para. 2.
[3] *Annotated Stat. 1915*, sec. 3521.

In Ohio, provision has been made for the periodic examination of workers exposed to lead poisoning, the employer being required to provide for the medical service. The requirements of this law are important evidences of the need for similar regulations in other states. The law, in part, reads as follows:

The employer shall cause every employee who, while engaged in any work or process is exposed to lead dusts, lead fumes or lead solutions, to be examined at least once a month for the purpose of ascertaining if symptoms of lead poisoning appear in any employee. The employee shall submit himself to the monthly examination and to examination at such other times and places as he may reasonably be requested by the employer. The examinations shall be made by a licensed physician, designated and paid by the employer, and shall be made during the working hours, a time allowance therefor, at the employer's expense, being made to each employee so examined.

Every physician making an examination and finding what he believes to be symptoms of lead poisoning shall enter, in a book to be kept for that purpose in the office of the employer, a record of such examination. The examining physician shall also within forty-eight hours, report such examination and finding to the employer, and after five days from such report the employer shall not continue the said employee in any work or process where he will be exposed to lead dusts, lead fumes or lead solutions, nor return the said employee to such work or process without a written permit from a licensed physician.[1]

New Jersey has also recognized the seriousness of lead poisoning, the law of the state noting that:

Every work or process in the manufacture of white lead, red lead, litharge, sugar of lead, arsenate of lead, lead chromate, lead sulphate, lead nitrate, or fluosilicate is hereby declared to be especially dangerous to the health of the employees who,

---

[1] *Ohio Gen. Code 1910,* sec. 6330.

while engaged in such work or process are exposed to lead dusts, lead fumes or lead solutions.

Every work or process in the manufacture of pottery, tiles or porcelain enameled sanitary ware is hereby declared to be especially dangerous to the health of the employees who, while engaged in such work or process, are exposed to lead dusts or lead solutions.

Every employer shall, without cost to the employees, provide the following devices, means and methods for the protection of his employees who, while engaged in any work or process, are exposed to lead dusts, lead fumes or lead solutions.

The employer shall provide and maintain workrooms adequately lighted and ventilated, and so arranged that there is a continuous and sufficient change of air, and all such rooms shall be fully separated by partition walls from all departments in which the work or process is of nondusty character; and all such rooms shall be provided with a floor permitting an easy removal of dust by wet methods or vacuum cleaning, and all such floors shall be cleaned either by wet method or vacuum cleaned daily.

Every work or process, including the corroding or oxidizing of lead, and the crushing, mixing, sifting, grinding, and packing of all lead salts or other compounds, shall be so conducted and such adequate devices provided and maintained by the employer as to protect the employee, as far as possible, from contact with lead dust or lead fumes.[1]

## REGULATION OF BAKERIES, FOOD ESTABLISHMENTS, ETC.

Laws regulating the conditions of employment in bakeries, confectioneries, ice-cream factories, dairies, canneries, and the like, combine the industrial factor with ideas of the protection of the public health.[2]   It follows from the nature of the industries affected that the welfare of the employee is frequently only a secondary aim in the enactment of the laws,

---

[1] *N. J. Act 1914*, secs. 2-3.

[2] See, for example, *Mass. Gen. Laws*, ch. 111, secs. 34-48; *N. Y. Acts of 1921*, ch. 50, secs. 330-333.

the restrictions laid upon him often being less for his own welfare than to secure the cleanliness and wholesomeness of the product.

The laws are varied but include some or all of the following provisions: that the establishment shall be properly lighted, drained, plumbed and ventilated; that screens must be provided for the doors and windows and the food and materials protected from flies, etc; that the clothing of all operatives must be clean and sanitary; that the interior walls must be washed, whitewashed, or painted at reasonable intervals; that suitable and separate toilets be kept for each sex and that they be maintained in a sanitary condition and that washing facilities be supplied and used; that cuspidors must be provided, the same to be cleaned daily; that smoking and expectoration be prohibited; that no one be allowed to use any workroom as a sleeping room; that no persons suffering from any communicable disease be employed; and that employees with infected wounds be prohibited from handling food until the wound is healed.

The regulation of work in cigar factories might perhaps be included under this section. The Wisconsin law,[1] for instance, provides that:

Every room in which cigars are manufactured while work is carried on shall be so ventilated that the air shall not become impure and injurious to the health of the persons employed therein, and it shall wherever necessary, by the means of air shafts or other ventilation, be so changed as to render harmless all gases, dust, and other impurities generated in the process of manufacturing cigars. All windows are to be kept open for thirty minutes before working hours and for thirty minutes after working hours.

[1] *Stat. 1923*, sec. 110.04.

### CONTROL OF INFECTION FROM WIPING CLOTHS

It has been known for some time that workers using soiled rags for wiping machinery, printing presses, etc., have developed severe ailments which could be directly traced to their occupations. To control this situation, some of the states have passed regulatory legislation dealing with this matter. The Massachusetts law provides that: " All publishers and printers shall use a sanitary cloth or other sanitary material in cleaning their presses." [1] The Ohio law stipulates that wiping rags " shall be thoroughly washed with soap and alkali soda, sterilized with chemical preparations, and dried with an average heat of 212 degrees ".[2] The law of California goes into considerable detail, some elements of which are as follows:

Every person or corporation who supplies or furnishes to his or its employees for wiping rags, or who sells or offers for sale for wiping rags, any soiled wearing apparel, underclothing, bedding, or parts of soiled or used underclothing, wearing apparel, bed clothes, bedding or soiled rags and cloth, unless the same have been sterilized by a process of boiling for forty minutes in a solution containing five per cent of caustic soda, and unless before such boiling, the sleeves, legs and bodies of garments are ripped, and made into flat pieces, is guilty of a misdemeanor.

Wiping rags within the meaning of this act are cloths and rags used for wiping and cleaning the surfaces of machinery, tools, locomotives, engines, motor cars, automobiles, cars, carriages, windows, and furniture, and surfaces of articles, appliances and engines in factories, shops, steamships and steamboats, and generally used for cleaning purposes in industrial employments, and also used by mechanics and workmen for wiping from their hands and bodies soil incident to their employment.

Every package or parcel of wiping rags must, before being sold or offered for sale, be plainly marked " sterilized wiping

---

[1] *Gen. Laws 1921*, sec. 142.

[2] *Gen. Code 1910*, sec. 1011 (added 1923, p. 314).

rags," with the number and date of permit given for the conducting of the laundry in which the rags contained in such package or parcel were laundered and sterilized, and the name of the board or officer issuing the permit; or with the name and location of the laundry in which such rags were laundered and sterilized.[1]

### WORK IN COMPRESSED AIR

The increasing construction of tunnels, subways, bridges and sky-scrapers in all parts of the country is making it imperative that state control be exercised over the employment of workers under compressed air. However, but few of the states have thus far enacted legislation regulating employment in this hazardous occupation. The laws in New Jersey,[2] New York[3] and Pennsylvania[4] are quite detailed, and, it has been found, that when these provisions are followed, virtually no cases of caisson disease, or the "bends" are apt to develop. While states other than those mentioned have not enacted legislation to control this particularly painful occupational disease, the attempt has been made to control the situation through the method of administrative orders.[5] The several laws and orders generally include provisions for the physical examination of all applicants for work, and of all employees at stated intervals; a sliding scale of working hours, decreasing as the pressure increases, and a period of gradual "decompression", ranging from one minute for

[1] *Acts of 1913*, ch. 81, secs. 1-6.

[2] *Acts of 1914*, ch. 121. This law was prepared by the American Assn. for Labor Legislation and adopted without change. (See: *Amer. Labor Legis. Rev.*, vol. iv, 1914, p. 549.)

[3] *Consol. Laws 1909*, secs. 425-437. This was the earliest American statute designed to combat this hazard. The law has been amended and strengthened several times since it was first enacted.

[4] *Stat. 1920*, secs. 5424-5436.

[5] Commons, J. R. and Andrews, J. B., *Principles of Labor Legislation* (New York, 1920), p. 371.

emergence from a pressure of 10 pounds above normal to 25 minutes from emergence from a pressure of 50 pounds above normal.    Work under more than 50 pounds pressure is forbidden.    The employer must maintain dressing rooms with lockers, hot and cold shower baths, and provision for drying clothes.    Medical attendants are also required, as well as a hospital lock for the treatment of sufferers from caisson disease.

### REGULATION OF ANTHRAX

Mention has previously been made of the fact that one of the chief dangers to workmen in the handling of hides, skins, hair, wool, etc., is anthrax infection which may result from contact with any of these animal by-products.    In the United States the Department of Agriculture and the Treasury Department have given attention to the matter, and on October 15, 1917, issued a joint order effective January 1, 1918, giving regulations governing the sanitary handling and control of hides, sheepskins, goatskins, hair, wool and other animal by-products offered for entry into the United States.    The regulations promulgated require among other things, the disinfection of the imported materials, specifying the ways and means of disinfection prior to shipment, or after arrival in the United States.[1]

### REGULATION THROUGH INDUSTRIAL CODES

Labor legislation has within recent years reached a highly voluminous and technical stage.    The legislatures, particularly in large industrial states, cannot possibly consider all the facts and details relating to hazardous labor conditions. To meet this problem, the plan was developed of delegating legislative power to the administrative body enforcing the labor law.    In New York, for instance, the law provides that the rules of the industrial board may be made for:

[1] *U. S. Bur. Lab. Stat. Mo. Lab. Rev.*, vol. ix, no. 1, July, 1919, pp. 180-184.

Guarding against and minimizing fire hazards, personal injuries and diseases. . . . Whenever the board finds that any industry, trade, occupation, or process involves such elements of danger to the lives, health, or safety of persons employed . . . the board may make special rules to guard against such elements of danger by establishing requirements as to temperature, humidity, the removal of dusts, gases or fumes.

The rules of the board have the force and effect of law, and constitute the industrial code, which remains in force until amended or repealed.[1]  Under the provisions of this legislation, the industrial board in New York has established rules and regulations for the carrying on of a number of hazardous industries, which may be the cause of occupational diseases.  Among these are codes relating to: bakeries and confectioneries;[2] equipment, maintenance and sanitation of foundries and the employment of women in core rooms;[3] milling industry and malthouse elevators;[4] removal of dust, gases and fumes;[5] mines and quarries;[6] construction, equipment, maintenance and operation of laundries;[7] dry dyeing plants and dry cleaning plants.[8]

[1] *Consol. Laws 1909*, art. ii, secs. 28-29.

[2] N. Y. State Dept. of Labor, *Industrial Code Bull. no. 4*, effective June 15, 1914.

[3] N. Y. State Dept. of Labor, *Industrial Code Bull. no. 10*, effective April 15, 1915.

[4] N. Y. State Dept. of Labor, *Industrial Code Bull. no. 11*, effective April 15, 1915.

[5] N. Y. State Dept. of Labor, *Industrial Code Bull. no. 12*, effective May 15, 1915.

[6] N. Y. State Dept. of Labor, *Industrial Code Bull. no. 27*, effective Oct. 1, 1924.

[7] N. Y. State Dept. of Labor, *Industrial Code Bull. no. 17*, effective July 1, 1918.

[8] N. Y. State Dept. of Labor, *Industrial Code Bull. no. 29*, effective Nov. 1, 1926.

An important example of this method of controlling occupational diseases, without having recourse to state legislatures and the considerable delay usually involved in such a step, relates to the matter of spray painting. This process of applying paint has of recent years been spreading very rapidly with the consequent intensification of the already recognized hazards involved in the use of materials containing lead, benzol, or other toxic substance. Massachusetts,[1] Michigan [2] and Wisconsin [3] control the process of spray coating by a code of special rules and regulations. In California a series of spray-coating safety orders were issued and became effective November 1927; in Pennsylvania spray coating regulations were issued in 1928, while in New York a spray-coating code was still in the hands of an advisory committee on March 15, 1931. Other states are attempting to apply their general laws regarding adequate ventilation in dealing with the problem of spraying paint, or are expecting that the employer will adhere to the provisions of compensation acts obligating the employer to do what is reasonably necessary to protect the life, health, safety and wel-

[1] *Mass. Revised Rules and Regulations Pertaining to the Painting Business*, December, 1925. Rule 6—" Health Requirements", among which is the following: " Respirators or devices shall be furnished by the employer and kept in a sanitary condition by the employee ".

[2] Michigan, "Rules and Standards on Spray Coating of Manufactured or Fabricated Articles—Paints, Varnishes, Lacquers, Enamels, Stains, and Similar Surface Coatings, by Means of Compressed Air, etc., 1927." Rules 4, 6 and 7 make certain requirements regarding the design, construction and maintenance of booths in which the painting is carried on. Rule 11 requires the employer to provide respirators or other equally efficient protective devices when the material being sprayed is known to contain an ingredient which is injurious to health.

[3] Wisconsin, "General Orders on Spray Painting." Order 2057 requires the provision of effective respirators or other devices. Order 2060 deals with the use of cabinet booths, room booths, and the use of protective clothing covering the entire body except the face and neck, which in turn are offered protection by the respirator.

fare of workmen. The occupational hazards to which workers are exposed in spraying paint have only come to the fore in recent years, and adequate measures to cope with these hazards have not as yet been uniformly evolved. In the interim, the exposure to lead, benzol and other toxic substances continues without proper protection or compensation being provided for many of the workers.

### INDUSTRIAL MEDICAL WORK

The increased occupational hazards brought about by the industrial transformation of the nineteenth century paved the way for the introduction of the industrial physician. Today, his work is coming to be recognized as an integral part of the productive process in nearly all large scale production.

The primary stimulus to the introduction of medical service into industry has been the increasing pressure upon the employer through the enactment of factory inspection and workmen's compensation laws. These laws have placed upon the employer practically the entire responsibility for the safeguarding of his employees against accident, and occasionally against occupational disease. The enactment of compensation laws for the first time placed a definite economic value upon the loss which ensued from injuries and some diseases suffered by workmen in various types of employment. The employer thus knows beforehand the probable cost of such injuries and diseases, and has an incentive to reduce these costs by giving proper attention to accident and disease prevention, to the handling of the materials used, and to the health of the worker. It is now generally realized that only by careful attention to these factors can injury to, and illness of, workers be prevented. Consequently, employers in states having compensation laws have found it economically profitable to provide some form of medical supervision for employees.[1] In this way the employer has introduced a type of regulation for the protection of the worker. Such regulation has assumed several forms, including pre-employment medical examination to exclude such as might be harmed by exposure in certain occupations or to particular substances,[2] periodic examination during employment, study of causes of sickness [3] and absenteeism,[4] and other measures designed either to regulate or eliminate the development of occupational diseases.[5]

### PROHIBITION AS A METHOD OF CONTROL

The outstanding example of the application of the method of prohibition to a dangerous substance is the world-wide banishing of poisonous phosphorus from the match industry. Within eleven years after the commercial introduction of the phosphorus match in 1827 the disease known as " phossy jaw ", or phosphorus necrosis, was attracting the attention of government investigators in different countries. Finland, the first of all countries, in 1872 forbade the use of white phosphorus in match factories. Similar action was taken by Denmark in 1874, and by France in 1897. Other countries followed with the result that in 1906 the Interna-

1 National Industrial Conf. Bd., *Medical Care of Industrial Workers* (New York, 1926), p. 6.

2 Hackett, J. D., *Health Maintenance in Industry*, 1925, p. 226.

3 Brundage, D. K., *Disabling Sickness among Employees of a Rubber Manufacturing Establishment* (B. F. Goodrich Co., Akron, O.), U. S. Pub. Health Rept., vol. 37, no. 50, Dec. 15, 1922.

4 Brundage, D. K., *Sickness and Absenteeism During 1919 in a Large Industrial Establishment*, U. S. Pub. Health Rept., vol. 35, no. 37, Sept. 10, 1920.

5 In England, through the cooperation of large employers of labor, there has been organized the Industrial Health Education Society for the purpose of informing industrial workers as to the occupational diseases and disabilities to which they are liable in the particular trades, and methods of prevention. *The Lancet*, no. 5541 (London, Nov. 9, 1929), p. 989.

tional Association for Labor Legislation secured an international conference at Berne which later resulted in a treaty signed by a number of countries absolutely prohibiting the manufacture, importation or sale of matches made from white phosphorus.[1]

In the United States an investigation of phosphorus poisoning in the match industry was begun in 1909, and the findings published in 1910.[2] Following this report, President Taft became interested in the subject, and in his message to Congress on December 6, 1910 stated:

I invite attention to the very serious injury caused to all those who are engaged in the manufacture of phosphorus matches. The diseases incident to this are frightful, and as matches can be made from other materials entirely innocuous, I believe that the injurious manufacture could be discouraged, and ought to be discouraged, by the imposition of a heavy Federal tax. I recommend the adoption of this method of stamping out a very serious abuse.[3]

As a result, the Esch-Hughes bill was introduced in Congress, passed, and signed by President Taft on April 9, 1912, to go into effect July 1, 1913.[4] This law required manufacturers using the common poisonous white or yellow phosphorus to pack such matches in prescribed packages and pay a tax of two cents per 100 matches. Further, the importation and exportation of phosphorus matches was absolutely forbidden, with provision for heavy penalties for violation of the law.[5] The result of this taxation was to render it prohibitive to manufacture such phosphorus matches in compe-

---

[1] Commons and Andrew, *op. cit.*, pp. 354-355.

[2] Andrews, J. B., "Phosphorus Poisoning in the Match Industry," *U. S. Bur. Lab. Bull. no. 86*, Jan., 1910, pp. 31-146.

[3] *Amer. Labor. Legis. Rev.*, vol. i, 1911, p. 98.

[4] *Ibid.*, vol. iv, 1914, p. 534.

[5] *U. S. Stat.*, Act of April 9, 1912, secs. 6271-6287.

tition with so-called safety matches, thus eliminating at one stroke the cases of phosphorus necrosis arising in the match industry.[1]

A situation similar to that of phosphorus matches came to the attention of the United States Department of Labor, through the discovery of a number of cases of phosphorus necrosis among those who had worked in factories making fireworks. A careful study was made of the subject, resulting in the recommendation that the use of white or yellow phosphorus in the manufacture of fireworks should be prohibited.[2] Through the offices of the United States Bureau of Labor Statistics, an agreement was reached with all the manufacturers of fireworks employing white or yellow phosphorus that they would cease so to manufacture on or before August 15, 1926. Part of this agreement, which is a novel form of prohibition by agreement, reads as follows:[3]

We, the undersigned manufacturers of the articles hereinafter named, agree jointly and each on his own behalf that we will discontinue on or before the 15th of August 1926, the manufacture of any type, form, or style of fireworks containing white or yellow phosphorus, and that after the disposal of the present stocks on hand and specifically after April 1, 1927, we will not

[1] To the American Assn. for Labor Legislation and its officers belongs much of the credit for bringing about the desired result. The Association was successful in inducing the Diamond Match Co. to assign its patent for one of the most available substitutes for the poison to three trustees, Prof. Seligman of Columbia University, Attorney Ralston of the American Federation of Labor, and Commissioner Neill of the U. S. Bureau of Labor. As even this extraordinary step was not sufficient to entirely allay suspicion of monopoly, the owners at the request of the trustees, concurred in by President Taft, cancelled the patent on Jan. 28, 1911. See *Amer. Labor Legis. Rev.*, vol. i, 1911, p. 98.

[2] *U. S. Bur. Lab. Stat. Bull. no. 405*, "Phosphorus Necrosis in the Manufacture of Fireworks and in the Preparation of Phosphorus," 1926, p. 5.

[3] *Annual Report*, U. S. Secretary of Labor, 1927, pp. 41-42.

sell or offer for sale any forms of fireworks, novelties, or products or other devices that contain white or yellow phosphorus.

And we hereby agree with the Secretary of Labor not only to cease the manufacture and sale of these articles on the dates hereinbefore specified, but to agree to any form of legislation or rules or regulations which may be instituted to prevent others from engaging in the manufacture or sale of such commodities, believing as we do that the injury resulting from such articles far exceeds their worth to the public.

### INTERNATIONAL PROTECTION OF LABOR

Within recent years there has been a growing realization of the desirability of international action in the protection of workers in various countries against occupational diseases. Mention has already been made of the work of the International Association for Labor Legislation, and the agreement in the case of phosphorus used in the manufacture of matches. Those closely allied with the work of this Association, which held its first sessions in Paris in 1900, realized that, like other private associations, its work at best could only be along propaganda lines and without official weight. However, the International Association, with permanent headquarters in a government building at Basle, Switzerland, has been in receipt of subventions from twenty-two different national governments. With the organization of the League of Nations, certain of the activities of the International Associations were taken over by the International Labour Office, which was instituted in association with the League.

In the Treaty of Peace with Germany and the Covenant of the League of Nations, provision was made for attention to be paid to " The protection of the worker against sickness, disease and injury arising out of his employment ".[1]

---

[1] Part xiii, sec. 1.

Subsequently, the International Labour Organization took action on several well-known occupational diseases. At an early meeting of the International Labour Conference it was recommended that each member of the International Labour Organization, which had not already done so, should adhere to the International Convention adopted at Berne in 1906 in the prohibition of the use of white phosphorus in the manufacture of matches. A wider acceptance of the Berne Convention has resulted.[1]  Regarding anthrax, a recommendation was made and adopted that: " Arrangements should be made for the disinfection of wool infected with anthrax spores, either in the country exporting such wool, or, if that is not practicable, at the port of entry in the country importing such wool." [2]  The General Conference of the Organization made definite recommendations regarding the employment and protection of women and children in certain processes utilizing lead.[3]  A " convention " (No. 14) concerning the use of white lead in painting, adopted by the Conference on November 19, 1921, entered into force on August 31, 1923. The ratifying states undertook to prohibit the use of white lead and sulphate of lead and of all products containing these pigments, in the internal painting of buildings. The employment of males under 18 years of age and of all females in painting work involving the use of the barred pigments was prohibited.[4]

While international action is most desirable, still it remains for individual governments to agree to recommendations of the General Conference, which are usually arrived at after considerable study. As the United States is not a member

[1] *Record of the Inter. Labour Organization*, 1919-1928, World Peace Foundation (Boston, 1928), p. 128.

[2] *Ibid.*, pp. 129-130.

[3] *Ibid.*, p. 129.

[4] *Ibid.*, p. 128.

of the League of Nations nor of the International Labour Organization we are deprived of whatever benefit might accrue to labor in this country through such participation.

### CONCLUSION

The preceding summary of laws and codes regulating employments, having as their aim the protection of the health and lives of the workers, and the prevention of occupational diseases, indicates clearly that law-makers and others in this country are aware of some of the occupational diseases to which the workers are exposed. Likewise, the unusual action which seemed warranted in the case of phosphorus poisoning was but another index of the same thing. Unfortunately, the scope of the various regulatory laws and codes in many cases far exceeds the provision made for their enforcement, which requires a large corps of trained inspectors and others expert in special fields, and often involves difficulty or delay in imposing penalties for infringement. Further, the difficulty with all controlling and regulating legislation is to adapt it practically and intelligently to the constantly varying needs of rapidly growing industry, and the health problems that arise through the use of newly developed substances or in totally new processes. In order to meet these difficulties, constant study and research become necessary. The result is that before any new regulations or prohibitions can be developed and applied, many workers are exposed to the hazards of occupational diseases, without, frequently, any recourse whatsoever.

Careful studies have indicated that workers suffer from numerous occupational diseases which do not directly cause death. Secondary and terminal diseases often occur, which cover up the condition that originally caused or accelerated the death.[1] This matter is frequently lost sight of when tab-

[1] *U. S. Bur. Lab. Stat. Bull. no. 207*, March, 1917, p. 84.

ulating the occupational diseases which might have resulted in mortalities. It is, of course, apparent that regulation of industries and industrial processes which expose the worker to occupational diseases, and which render him liable to many types of illness of often undefined origin, must be continued and expanded as time goes on. Also, that adequate administrative and research staffs must be provided to carry on the regulatory and preventive activities. How to care for the workers who develop occupational diseases even before proper regulatory and preventive action can be taken, remains a serious problem.

The extent of occupational diseases in dusty, metal, chemical and miscellaneous industries even after numerous regulatory laws and codes have been in effect for many years, is undoubtedly an index of the need for some method of lessening the burdens which rest upon the industrial workers. Workmen's compensation legislation was enacted to help workers who suffered industrial injuries. Subsequently compensation for certain occupational diseases was provided in some states through extension of the workmen's compensation provisions. Despite this advance, as will be indicated, many injustices result to those who develop occupational diseases. It is for these reasons that some kind of health insurance to cover occupational diseases seems to many to be desirable.

# CHAPTER V

## WORKMEN'S COMPENSATION LEGISLATION AND JUDICIAL DECISIONS RELATING TO OCCUPATIONAL DISEASES

### INTRODUCTION

Compensation to the injured workmen is based upon the theory that the consumer of economic goods should bear all the expenses incurred in the production of such goods. Among these expenses must be included the pecuniary losses from deaths and injuries, as well as occupational diseases contracted in the regular course of production. Wages lost, medical attendance and burial expenses, in case of accidental injury or death, are all losses which should be considered as a part of the expense of production. If these losses are borne by the workman, he indirectly carries part of the expense of production. In order to avoid this, the expense of work-accidents should seemingly be considered as a part of the production costs.[1] By placing upon the employer the partial cost of work accidents and in some instances, of occupational diseases as well, a method of regulation is exercised which functions to check in part at least the industrial accident and disease rates.

The movement toward compensation for occupational diseases followed inevitably the scheme of accident compensation, for it early became obvious that elementary justice required the extension of similar relief to the victims of specific industrial diseases contracted in the course of employ-

[1] Commons and Andrews, *Principles of Labor Legislation*, p. 386.

ment. Though great difficulties were and are being encountered in the application of workmen's compensation laws for injuries arising out of employment, it has been found that the utilization of similar measures as compensation for occupational diseases, involves in many instances even greater problems. This may call for the application of methods other than workmen's compensation laws for the timely and adequate relief of workers injured through occupational diseases. However, before entering upon a discussion of the compensation laws in their application to occupational diseases, brief mention will be made of the development of workmen's compensation for accidental injuries arising out of and in the course of employment.

## WORKMEN'S COMPENSATION IN GERMANY

Workmen's compensation has been defined as a system of providing for partial, prompt and definite relief against disability or death resulting to workmen from industrial accidents (or diseases), without regard to questions of negligence. It involves the provision of medical, surgical, hospital and rehabilitation service, and payment of benefits based upon a fixed percentage of the injured workman's earnings at the time of the accident. It is designed primarily to mitigate the economic shock which invariably results from an industrial accident.[1]

Germany was the first country to adopt a system of compulsory insurance of workmen on a national scale. This policy was not regarded in Germany as a break with previous traditions, but was considered rather as a logical development of institutions for the care of disabled workmen made necessary by the change in conditions brought about by modern industrial methods. The beginnings of workmen's in-

[1] Harper, Samuel, A., *The Law of Workmen's Compensation* (Chicago, 1920), 2nd ed., p. 3.

surance in the area comprised within the former German Empire in many cases date back as far as the 18th century, and in a few instances to an even earlier date. The first forms of such insurance are found in the provisions for the relief of disabled miners, seamen and domestic workers. In the case of miners and seamen, it will be noticed, that these were the industries in which the safety of the workman was peculiarly dependent on the general management, and in which the risk of accident and sickness was accentuated because of such dependence on the care of fellow-workmen in the exercise of their duties, or on the skill and care of the one directing the workman's operations. Even in the 18th century, therefore, those industries which possessed some of the characteristics of modern industrial methods had developed special institutions for the relief of the workmen engaged in them.[1]

Under the German common law a workman injured in the course of his employment through the fault of the employer had to secure redress by means of a suit against the latter, subject to all the disadvantages which were recognized in Germany even in the early part of the 19th century. To meet the new conditions of industry, as well as to make provision for disabled workmen without means, many of the local governments throughout Germany enacted laws requiring specified workers within their areas to pay regular contributions to the communal treasury from which was paid sickness and accident relief. These steps antedated the national compulsory workmen's insurance by ten to twenty years, and the operation of these funds formed an important

---

[1] An Act adopted by the Reichstag on December 16, 1927 makes sickness insurance compulsory for seamen, but at the same time maintains certain of the ship owner's duties, especially while the seaman is on board. For further particulars, see: *Industrial and Labor Information*, vol. xxv, no. 5, Jan. 30, 1928, Inter. Labour Office (Geneva), pp. 158-161.

argument for the institution of a similar system of provision on a national scale. It may be said, therefore, that the national system of workmen's compensation insurance in Germany was adopted only after a thorough trial of other methods had shown that they were inadequate to meet the problem. No proposal to restrict the plan or to return to the former liability system has ever been seriously offered.[1]

The first compensation law was enacted on July 6, 1884. This law was amended and extended from time to time, until the year 1911, when all the provisions of the various acts were unified and codified into an act which was intended to take effect January 1, 1912. Owing to the administrative changes involved in the consolidation of the various parts of the social insurance laws, it did not take effect until January 1, 1913.

The compensation laws in Germany are divided into three general divisions: (a) sickness insurance; (b) accident insurance and (c) disability insurance. Contributions by the workmen themselves play an important part in the system. These insurance funds are managed by officers elected by workmen and employers. The accident fund is supported entirely by contributions from the employers and is managed by them. All employers are required to join the accident insurance fund of the trade in which they participate as a condition to engaging in the particular trade. There is governmental supervision of the entire system.[2]

The effects of the war and post-war periods have been to awaken public interest much more fully than ever before to the necessity of an organized system of health protection as the basis for a broad health policy. The text of Article 161

---

[1] U. S. Commissioner of Labor, *24th Ann. Report*, 1909, " Workmen's Insurance and Compensation Systems in Europe," vol. i, pp. 977-978.

[2] *Social Insurance Code of 1911*, pt. iii. This code was revised in 1924, but its origin goes back to the period 1883 to 1889.

of the Constitution of the Reich of 1919 expresses the new conception as follows: " The Reich shall create a comprehensive insurance system, in the conduct of which the persons insured shall play a leading part for the preservation of health and working capacity." [1]

The legal position of German insurance at present, is regulated by three important and separate acts, which provide for five different authorities and organizations. Sickness, invalidity and old-age insurance are organized on the basis of contracts of employment; insurance against industrial accidents in membership of certain trades or vocations. These types of insurance, however, do not cover all the employee classes of the population, but are supplemented by special acts for the benefit of employees. [2]

### WORKMEN'S COMPENSATION IN GREAT BRITAIN

Until 1880, in Great Britain, the employer was held liable by common law only for damages resulting from accidents directly attributable to his negligence or wrongful act. The Workmen's Compensation Act of 1897 fundamentally changed the relation between employer and workman in the occupations it covered. Under this act, unless the employee was guilty of a gross fault, the employer had to provide compensation to the workman injured in the ordinary course of his employment. The workman was no longer required to prove negligence on the part of the employer—it was enough that he was injured in the course of his employment. The Act of 1897 applied only to employments on or in or about a

---

[1] Goldmann, Franz and Grotjahn, Alfred, "Benefits of the German Sickness Insurance System," *Inter. Labour Office Studies and Reports,* Series M., no. 8 (Geneva, 1928), p. 1.

[2] For statistics on membership, benefits, etc. in the various organizations, see *Statistic des Deutschen Reichs: Band 338,* Die Krankenversicherung (Berlin, 1927).

railway, factory, mine, quarry, engineering work or work on any building exceeding thirty feet in height.[1]

The Workmen's Compensation Act of 1906 applied virtually to all industrial occupations, including domestic servants. Under this act every workman who suffers personal injury " arising out of and in the course of the employment " is entitled to compensation. The act specifically states that it does not apply to non-manual workmen, and to a few other groups.[2] The act also provides that a workman who suffers personal injury as a result of one of the 31 industrial diseases listed in the act is similarly entitled to compensation. An interesting feature relating to claims for compensation arising out of occupational diseases, is the provision that the workman is bound to supply the employer with the names and addresses of all other employers who employed him during the preceding twelve months. This is to enable the employer against whom the claim is made to obtain a contribution from previous employers whose employment may have had some relation to the setting-up of the disease. If when commencing fresh work a workman who has in a previous employment suffered from a disease, wilfully and falsely represents himself in writing as not having previously suffered from the disease, any claim against the new employer is barred.[3]

### DEVELOPMENT IN THE UNITED STATES

The form of legislation in the United States has been strongly influenced by the British Compensation Acts of 1897 and 1906. Originally, the American laws followed, to a considerable extent, the British statute in placing a greater

[1] Cohen, Joseph L., *Workmen's Compensation in Great Britain* (London, 1923), p. 93.

[2] *Ibid.*, p. 97.

[3] Thompson, W. H., *Workmen's Compensation* (London, 1922), p. 42.

obligation on the employers. Shortly after the laws were passed in this country, requirements were inserted compelling employers to insure their workmen's compensation liability, as a better security to workmen and the dependents of those who were killed, in the cases of long deferred payments. In affecting this insurance there were created in some of the states what have come to be generally known as " State Insurance Funds." Properly speaking, they are state managed funds, though they are not managed in the sense that the German funds are, nor do they have the governmental authority which is back of the British national insurance plan. They are merely administered by state officials without state guaranty, and employers have little or no voice in such administration.[1]

While fundamentally affected by foreign examples, the legislation in the United States is distinctly modified by local traditions, phraseology and various political and social influences. Constitutional limitations have likewise produced many dissimilarities in detail and phraseology, and in the consequent judicial and administrative construction. The resultant obligation of the employer to assure limited compensation for work injuries, and the reciprocal duty and right of the employee to accept it, vary greatly in the different states.[2]

## THE SCOPE OF WORKMEN'S COMPENSATION

The proper administration of workmen's compensation acts necessitates an appreciation of the legislative purpose to abolish the common law system relating to injuries to em-

---

[1] See for example, *N. Y. State Workmen's Compensation Law*, art. v, secs. 90-91, stating: " The fund shall be administered by the industrial commissioner," and " The state treasurer shall be the custodian of the state insurance fund."

[2] *Workmen's Compensation Acts in the United States*, National Industrial Conference Board, 1919, p. 2.

ployees as inadequate to meet modern conditions and conceptions of moral obligations, and substitute therefore a system based on a high conception of man's obligation to his fellow man.   The principle which was first advocated by Bismarck, and later by Lord Salisbury, is that in a modern industrial state the risk of injury to workmen while engaged in the employer's service is a social risk, chargeable against the business itself, the losses arising from which are to be added to the productive cost and to be borne ultimately by the community at large.   It permits an injured workman, or in the event of his death, his dependents, to demand as a right that which they were often compelled to ask as a charity, with ultimate costs in either event upon the community.[1]

The change made by this legislation is quite radical, and works fundamental changes in the familiar principles governing the employer's heretofore existing liability for negligence.   In place of the liability in an action for damages, in which the employer was liable only in case he or his representative was negligent or at fault, a liability is imposed on the employer for any accidental injuries to his employees arising out of the employment.   This is a liability which, as a general rule, is not conditioned on the employer's negligence or the employee's want of negligence.[2]

### ORIGIN AND NATURE OF COMMON LAW DEFENSES

The doctrines of assumption of risk, contributory negligence and negligence of fellow servant, have played an important part in British and American law.   Up to a very recent time, the rule had long been established that a workman

---

[1] Honnold, A. B., *Treatise on the American and English Workmen's Compensation Acts* (Kansas City, 1917), vol. i, p. 5.

[2] Compensation for injury regardless of fault, is the basis of the various workmen's compensation laws.   See: Appeal of Hotel Bond Co., 89 Conn. 143, 93 Atl. 245.

assumed all the risks which were necessarily incident to his employment; and also, all the risks which were obvious and concerning which he had actual information, or should have had knowledge by reason of the fact that they were open and obvious. This doctrine of assumption of risk was carried to the extent that even though the employer was negligent and failed in many respects to perform his duty in safeguarding his workmen, nevertheless if an employee continued to work after these risks, due to the negligence of the employer, had become obvious and well-known to the workman, he assumed such risk and could not recover if he was injured by reason of any of the risks thus assumed.[1]

The fellow-servant doctrine dates from 1837, when it was first established in Great Britain.[2] Before this year there is no record of any case in which it was sought to make an employer responsible for an injury incurred in his service, except in those cases where the injury could be traced directly or indirectly to his own personal negligence or breach of duty. In Priestly v. Fowler, the servant Priestly was employed by Fowler, and was riding in a van which was not under his control. Because of being overloaded by the negligence of another fellow-servant, the van broke down and injured Priestly. He sought to make his master responsible for such negligence but failed, the court ruling that:

The mere relation of the master and the servant can never imply an obligation on the part of the master to take more care of the servant than he may reasonably be expected to take care of himself. He is no doubt bound to provide for the safety of his servant, in the course of his employment, to the best of his judgement, information and belief. The servant is not bound to risk his safety in the service of his master, and may decline any

[1] Bradbury, Harry B., *Workmen's Compensation Law*, 3rd ed. (New York, 1917), pp. 18-19.

[2] Case of Priestly *v.* Fowler, 3 Meeson and Welsby 1, 6 England 1837.

service in which he reasonably apprehends injury to himself.
. . .[1]

In 1842 the defense of assumption of risk was first declared in this country as a principle involved in the application of the law of master and servant. It was stated that by virtue of an implied condition in the contract of service, the servant took upon himself the risks attending the performance of the service which he had engaged to perform, including injuries which might befall him from the negligence of his co-servants.[2] In a subsequent case[3] it was determined in substance, that a servant having knowledge of the circumstances attending the particular employment, in entering upon it, assumed the risk of injury that might befall him. From the reasoning and decisions in the preceding two cases developed, to a very large extent, the doctrine established on this subject in the different states.[4]

### ABROGATION OF COMMON LAW DEFENSES

No incident accompanying the subversion of the common law system of liability has attracted so much attention or been so frequently contested in court as the abrogation of the customary defenses of contributory negligence, assumption of risk and fellow-servant. The compulsion which the state has exercised by abolishing these defenses or interposing them to secure from employer or employee the acceptance of

[1] For a fuller discussion of this subject, see: Dawbarn, M. A., *Employers' Liability to their Servants at Common Law and under the Employers' Liability Act 1800*, 4th ed. (London, 1911).

[2] Farwell *v.* Boston and Worcester R. Co., 4 Metcalf (Mass.) 49 (1842). This was an action of an engineer who sued for damages because he had lost a leg due to the neglect of a switchman in failing to change a switch.

[3] Hayden *v.* Smithville Mfg. Co., 29 Conn. 548 (1861).

[4] Bailey, W. F., *A Treatise on the Law of Personal Injuries*, 2nd ed. (Chicago, 1912), vol. ii, pp. 939-940.

the new policy it desired to promote, is now generally regarded judicially as a justified form of what may be termed constitutional coercion.[1]

The issue as the courts have seen it, is not whether all rules of liability between employer and employee may be abrogated, but whether a reasonable body of new rules fitting the circumstances of modern industry may be substituted for the old. An arbitrary attempt to deprive the employer of his position under the old system without any beneficial status under the new was invalidated in a decision by the Supreme Court of Montana, in a case involving the maintenance of a state cooperative insurance fund for miners and laborers in and about mines.[2]

Many of the so-called common-law defenses were greatly modified, and in some instances, entirely abolished, before the compensation statutes were enacted. The alternative abolition of these defenses in the compensation acts was hit upon as a plan to escape the constitutional question in the case of Ives v. South Buffalo R. Co.[3] In this case it was held that a mandatory compensation law was unconstitutional, on the ground that it imposed liability on an employer independent of any act of negligence or default on his part, and therefore deprived him of his property without due process of law. In accordance with the suggestion of the decision in this case, a constitutional amendment enlarging the power of the legislature and enabling it to deal with the subject was

---

[1] Among the cases illustrating this, see: State *ex rel.* Yaple *v.* Creamer, 85 Ohio State 349 (1912); Borgnis *v.* Falk Co., 147 Wis. 327, 133 N. W. 209 (1911); Western Metal Supply Co. *v.* Pillsbury, 172 Cal. 407, 155 Pac. 491 (1915); Jenson *v.* So. Pacific Co., 215 N. Y. 514, 109 N. E. 600 (1915).

[2] Cunningham *v.* Northwestern Improvement Co., 44 Mont. 180, 119 Pac. 554 (1911).

[3] Ives *v.* So. Buffalo R. Co., 201 N. Y. 271, 94 N. E. 431 (1911).

adopted, and the subsequent enactment thereunder has been sustained.[1]

The viewpoint of the judiciary in vindicating the exchange of new regulation for old is summarized in the decision affirming the validity of the present New York Act:[2]

No one doubts that the doctrine of assumption of risk and the fellow-servant doctrine also, developed by the courts under different conditions than those now prevailing, may be limited and entirely abrogated by the legislature. . . . At common law the servant was held to assume by implied contract the ordinary risks of employment, including the risk of a fellow-servant's negligence, and even negligence imputable to the master. . . . It would not be a great extension of that doctrine for the legislature to provide that the employee should assume the risk of accidental injuries, and if that can be done it is certainly competent for the legislature to provide by the creation of an insurance fund for the limited compensation of the employees for all accidental injuries regardless of whether there was a cause of action for them at common law.

The legislature of New Jersey, after the decision in the Ives case, passed an optional compensation statute under which, if an employer failed to adopt the compensation principle in actions against him for personal injuries due to negligence, the common law defenses of assumption of risk and negligence of fellow-servant were abolished and the defense of contributory negligence was greatly modified. Other states followed the example of New Jersey.[3]

CONSTITUTIONALITY OF ELECTIVE AND COMPULSORY ACTS

The motives behind the legislation in the several states have been various and the legal theory upon which they have

[1] *Constitution of New York*, art. i, sec. 19, adopted Nov. 4, 1913.

[2] Jenson v. So. Pacific Co., 215 N. Y. 514, 109 N. E. 600 (1915).

[3] Bradbury, *op. cit.*, p. 21.

been based has been revolutionary, because the element of blameworthiness has been substantially ignored. Despite this, the legislation has been held constitutional.[1] Although compensation laws have been frequently attacked in the courts on the ground that they deprive the injured man and his dependents of the right of trial by jury, that they seek to confer judicial powers upon administrative bodies, that they violate the " due process " clause of the constitution, that the classification of industries is arbitrary and unreasonable, and that they deprive employers of common law defenses, they have generally been upheld by both federal and state courts.[2] There are only three states in which the first law enacted was held unconstitutional.[3]

The following quotation from the opinion of the United States Supreme Court in the case of New York Central R. Co. v. White is indicative of the views of the Court on the constitutionality of compensation acts generally.[4]

In considering the constitutional question, it is necessary to view the matter from the standpoint of the employee as well as from that of the employer. . . . The close relation of the rules governing responsibility as between employer and employee to the fundamental rights of liberty and property, is of course, recognized. But those rules, as guides of conduct, are not beyond alteration by legislation in the public interest. No person has a

[1] Clark, George L., *The Law of Torts*, 1926, pp. 256-257.

[2] Hawkins *v.* Bleakly, 243 U. S. 210, 37 Sup. Ct. 255 (1917); N. Y. Central R. Co. *v.* White, 243 U. S. 188, 37 Sup. Ct. 247 (1917); Crooks *v.* Tazewell Coal Co., 263 Ill. 343, 105 N. E. 132 (1914), Mountain Timber Co. *v.* Washington, 243 U. S. 219, 37 Sup. Ct. 260 (1917).

[3] Ives *v.* So. Buffalo R. Co., on the ground that it imposed upon employers a liability without fault or contract; Cunningham *v.* No. Western Improvement Co., 44 Mont. 180, 119, Pac. 554 (1911) held it denied to employers equal protection of the law; Kentucky State Journal Co. *v.* Workmen's Compensation Board, 161 Ky. 562, 172 S. W. 674 (1914), on various grounds.

[4] N. Y. Central R. Co. *v.* White, 243 U. S. 188, 37 Sup. Ct. 247 (1917).

vested interest in any rule of law entitling him to insist that it shall remain unchanged for his benefit.[1] . . . The common law bases the employer's liability for injuries to the employee upon the ground of negligence; but negligence is merely the disregard of some duty imposed by law; and the nature and extent of the duty may be modified by legislation, with corresponding change in the test of negligence. Liability may be imposed for the consequence of a failure to comply with a statutory duty, irrespective of negligence in the ordinary sense; safety appliance acts being a familiar instance.

The immunity of the employer from responsibility to an employee for the negligence of a fellow employee is of comparatively recent origin, it being the product of the judicial conception that the probability of a fellow workman's negligence is one of the natural and ordinary risks of the occupation, assumed by the employee and presumably taken into account in fixing his wages. It needs no argument to show that such a rule is subject to modification or abrogation by a state upon proper occasion. The same may be said with respect to the general doctrine of assumption of risk. . . . So, also, with respect to contributory negligence. . . .

The statute under consideration sets aside one body of rules only to establish another system in its place. If the employee is no longer able to recover as much as before in case of being injured, through the employer's negligence, he is entitled to moderate compensation in all cases of injury, and has a certain and speedy remedy without the difficulty and expense of establishing or proving the amount of the damages. . . . If the employer is left without defenses respecting the question of fault, he at the same time is assured that the recovery is limited. . . .

In the case of Mountain Timber Co. vs. The State of Washington, in which a compulsory act was passed, and including a monopolistic-compulsory state insurance plan, the majority opinion of the United States Supreme Court sus-

[1] See: Munn *v.* Illinois, 94 U. S. 113 (1878) ; Martin *v.* Pittsburg and Lake Erie R. Co., 203 U. S. 284 (1906).

tained the decision of the State Court and the validity of the compensation act. The court held in effect that the establishment of a compensation system is a public concern, and to be administered through state agencies.

The state is at liberty, notwithstanding the 14th Amendment to disregard questions of fault in arranging a system of compensation for such injuries. We are unable to discern any ground in natural justice or fundamental right that prevents the state from imposing the entire burden upon the industries that occasion the losses. The act in effect puts these hazardous occupations in the category of dangerous agencies, and requires that the losses shall be reckoned as a part of the cost of industry, just like the payroll, the repair account, or any other item of cost.

The Iowa act, unlike that of New York or Washington, is elective in form. In discussing the constitutionality of that act, the United States Supreme Court, in the case of Hawkins v. Bleakley said:

Some of the applicant's objections are based upon the ground that the employer is subjected to a species of duress in order to compel him to accept the compensation features of the act, since it is provided that an employer rejecting these features shall not escape liability for personal injury sustained by an employee arising out of and in the usual course of employment. . . . It is clear that the employer has no vested right to have the so-called common law defenses perpetuated for his benefit, and that the 14th Amendment does not prevent a state from establishing a system of workmen's compensation without the consent of the employer.

The decision in the Ives case, holding the compulsory act of 1910 in New York invalid, had a profound effect upon subsequent legislation relating to the workmen's compensation principle in the various states.[1] A large number of

[1] Brandbury, *op. cit.*, pp. 592-593.

elective laws and several compulsory statutes were passed, all aiming to apply the principle of adequate protection of the worker when injured.[1]

### COMPENSATION FOR OCCUPATIONAL DISEASES

Though workmen's compensation laws originally concerned themselves only with mechanical injuries, such as cuts, broken bones or loss of members, it soon became obvious that elementary justice required the extension of similar relief to the victims of specific industrial diseases contracted in the course of employment.

There is a definite distinction between occupational disease and occupational accident. Some of the main points of difference are the following: occupational disease is anything but exceptional, being in fact the consequence of ordinary work; such disease is inevitable, to the extent that it accrues from the repetition of the same work, and is generally not the result of a single occurence; the inception is usually slow, insidious and usually difficult to determine. In these matters occupational diseases differ sharply from accidents.[2]

The term "accident" was unsatisfactorily defined in the first British Act of 1897.[3] In the case of Brinton's Limited v. Turvey the decision of the House of Lords held that an infection from anthrax in the wool handled was "personal injury by accident". However, it was expressly emphasized that the decision must not be regarded "as involving

---

[1] Following are some of the leading cases in the state courts sustaining the constitutionality of their respective acts against the objection that they denied to employer or employee due process of law. State *v.* Creamer, 85 Ohio 349 (1912) State *ex rel.* Davis-Smith Co. *v.* Clausen, 65 Wash. 156, 117 Pac. 1101 (1911); Sayles *v.* Foley, 38 R. I. 484, 96 Atl. 340 (1916); Anderson *v.* Carnegie Steel Co., 255 Penn. 33, 99 Atl. 215 (1916).

[2] For detailed list of differences, see: *Compensation for Occupational Diseases*, Inter. Labour Office, Series M., no. 3 (Geneva, 1925), p. 13.

[3] National Industrial Conf. Bd., *Workmen's Compensation Acts in the United States* (New York, 1919), p. 39.

the doctrine that all diseases caught by a workman in the course of his employment are to be regarded as accident ".[1]

Parliament in succeeding legislation in the Act of 1906 extended the principle of workmen's compensation to industrial diseases, which was the natural outcome of the previous Act of 1897 and the decisions under it. Once the principle was admitted that workmen should be compensated by their employers for injuries by accident, it appeared only consistent that injuries or incapacity caused by disease due to the employment should be included.[2] The Act of 1906 included for compensation a schedule of six of the commonest occupational maladies.[3] To make the compensation act more elastic to meet new conditions as they might arise, it was provided that such other occupational diseases might be included in the act, by proclamation of the Secretary of State, as should be found to arise from the conditions prevailing in the different industries.[4] A departmental committee was at once appointed, and upon their recommendation, 16 additional diseases were added to the list by order dated May 22, 1907.[5] Other diseases were added from time to time, until a total of 28 occupational diseases or injuries have come to be included in the schedule. Among the diseases listed are the following: arsenic poisoning; lead poisoning; poisoning by benzene; dinitrophenol; nitrous fumes; " dope "; tetrachlorethane; carbonbisulphide; nickel carbonyl; African boxwood; dermatitis produced by dust or liquids; skin cancers or ulceration, due to certain substances, such as tar, pitch, chrome; compressed air illness; cataract in glass work-

[1] Brinton's Limited *v.* Turvey, 1905 A. C. 230.

[2] Lawes, E. T., *The Law of Compensation for Industrial Disease* (London, 1909), p. 3.

[3] Commons and Andrews, *op. cit.,* p. 396.

[4] Bradbury, *op. cit.,* p. 8.

[5] Lawes, *op. cit.,* pp. 3-4.

ers; miner's nystagmus; glanders; anthrax; mercury poisoning and phosphorus poisoning.[1] Lead and arsenic poisoning appear twice and were inserted merely to widen the scope of occupations to which the act applies.[2]

### EFFECT OF BRITISH ACTS IN THE UNITED STATES

The inclusion of occupational diseases in compensation laws has often been discussed in the United States, though action has not always followed the British example. The distinction which is drawn in nearly all the compensation acts in this country between "accidental injuries" and "occupational diseases" is oftimes both illogical and unjust to the injuried workers. The principal argument in favor of workmen's compensation acts has been that each industry should carry the burden of its own injured workers. The man who breathes lead fumes or other dust in his workshop and becomes incapacitated, is, to a very much greater extent, the victim of the industry in which he is engaged, than, for example, the worker who falls on the stairs of his workplace on his way home. The latter will generally receive compensation, whereas the former frequently is denied the protection of the workmen's compensation law.

The popular conception of an "accident" is probably much narrower than the definition which that term now receives in the construction of compensation legislation. The fact is that the original conception has been greatly modified and extended by the adoption of broader statutory language and by administrative construction. The English compensation act, which has served as a model for much of the legislation in the United States, created its fundamental liability by the phrase "personal injury by accident arising out of and in the course of employment."[3]

[1] Thompson, *op. cit.*, pp. 88-91.

[2] Kober and Hayhurst, *op. cit.*, p. 1017.

[3] National Industrial Conf. Board, *Research Report*, no. 1, p. 38.

OCCUPATIONAL DISEASE COMPENSATION IN UNITED STATES

The construction and practice of administrative commissions indicates an increasing tendency to allow compensation for many forms of disease contracted during employment without requiring a clear proximate relation thereto to be established. The courts, however, appear to be drawing a distinction between industrial diseases and disease resulting from accident, allowing compensation in the latter class of cases and denying it in the former unless definitely included by the terms of the statute.

No law in its original enactment made specific provision for compensating occupational diseases. The dominant idea of accident has given way by degrees, however, until at present 12 states and the federal government provide compensation, either for occupational diseases generally or for designated diseases of this class.[1]

The laws may generally be classified into two types: those in which it is provided that awards shall be given in cases of compensable diseases generally, and those listing specific occupational diseases for which an award will be granted. Under the laws providing compensation for occupational diseases in general terms, it should be noted that the language of the acts is sometimes limited to one general class of diseases (Illinois and Kentucky) or limited to an injury (Massachusetts).

The following are the laws in effect in the states indicated and defining the terms "injury" and "occupational disease".

## California

The term "injury" as used in this act, shall include any injury or disease arising out of the employment including injuries to artificial members. In case of aggravation of any disease

---

[1] *U. S. Bur. Labor Stat. Bull. no. 423*, Dec., 1926, p. 63.

existing prior to such injury, compensation shall be allowed only for such proportion of the disability due to the aggravation of such prior disease as may reasonably be attributed to the injury.[1]

### Connecticut

The words "personal injury" or "injury" as the same are used shall be construed to include only accidental injury which may be definitely located as to the time when and the place where the accident occured, and occupational disease as herein defined. The words "occupational disease" shall mean a disease peculiar to the occupation in which the employee was engaged and due to causes in excess of the ordinary hazards of employment as such. The words "arising out of and in the course of his employment," shall mean an accidental injury happening to an employee or an occupational disease of such employee originating while he shall have been engaged in the line of his duty in the business or affairs of the employer upon the employer's premises, or while so engaged elsewhere upon the employer's business or affairs by the direction, express or implied, of the employer. A personal injury shall not be deemed to arise out of the employment unless casually traceable to the employment other than through weakened resistance or lowered vitality.[2]

### District of Columbia

The term "injury" means accidental injury or death arising out of and in the course of employment, and such occupational diseases or infection as arises naturally out of such employment, or as naturally or unavoidably results from such accidental injury, and includes an injury caused by the wilful act of a third person directed against an employee because of his employment.[3]

### Illinois

Every employer in this state engaged in the carrying on of any process of manufacture or labor in which sugar of lead,

[1] *Stats. 1917*, ch. 586, as amended 1919, ch. 471.

[2] *Gen. Stats. 1918*, sec. 5388, as amended 1927, ch. 307, sec. 7.

[3] 45 Stat. 600 and 44 Stat. 1424.

white lead, lead chromate, litharge, red lead, arsenate of lead, or Paris green are employed, used or handled, or the manufacture of brass or the smelting of lead or zinc which processes and employments are hereby declared to be especially dangerous to the health of the employees engaged in any process of manufacture or labor in which poisonous chemicals, minerals or other substances are used or handled by the employees therein in harmful quantities or under harmful conditions, shall provide for and place at the disposal of the employees engaged in any such process or manufacture and shall maintain in good condition and without cost to the employees, proper working clothing, etc., and in all processes of manufacture or labor referred to in this section which are unnecessarily productive of noxious or poisonous dusts, adequate and approved respirators shall be furnished, etc.[1]

If an employee is disabled or dies, and his disability or death is caused by an occupational disease arising out of and in the course of his employment in one or more of the occupations referred to in this act, he or his dependents, as the case may be, shall be entitled to compensation, in the same manner and subject to the same terms, conditions, and limitations as are now or may hereafter be provided by the workmen's compensation act for accidental injuries sustained by employees arising out of and in the course of their employment; and for this purpose the disablement of an employee by reason of an occupational disease, arising out of and in the course of his employment in one or more of the occupations referred to in this act, shall be treated as the happening of an accidental injury.[2]

### Kentucky

It shall affect the liability of the employers subject thereto to their employees for personal injuries sustained by the employee by accident arising out of and in the course of his employment, or for death resulting from such accidental injury; provided, how-

[1] *Rev. Stat. 1917*, ch. 48, sec. 154, p. 1469.
[2] *Acts of 1923*, p. 352.

ever, that personal injury by accident as herein defined shall not include diseases except where the disease is the natural and direct result of a traumatic injury by accident, nor shall they include the results of a preexisting disease but shall include injuries or death due to inhalation in mines of noxious gases or smoke, commonly known as " bad air ", and also shall include the injuries or death due to the inhalation of any kind of gas.[1]

### Massachusetts

If an employee * * * receives a personal injury arising out of and in the course of his employment, he shall be paid compensation. * * * [2]

### North Dakota

" Injury " means only an injury arising in the course of employment, including an injury caused by the wilful act of a third person directed against an employee because of his employment, but shall not include injuries caused by the employee's wilful intention to injure himself or to injure another. The term " injury " includes in addition to an injury by accident, any disease proximately caused by the employment.[3]

### Wisconsin

The provisions of sections 102.01 and 102.34, both inclusive, are extended so as to include, in addition to accidental injuries, all other injuries, including occupational diseases, growing out of and incidental to the employment.[4]

### Federal Civil Employees

The term " injury " includes, in addition to injury by accident, any disease proximately caused by the employment.[5]

[1] Acts of 1916, ch. 33, sec. 1, as amended 1918, ch. 176; 1922, ch. 50; 1924, ch. 70.

[2] Acts of 1919, ch. 162, sec. 2, as amended 1921, ch. 142; 1925, ch. 222.

[3] Gen. Laws 1921, ch. 152, sec. 26.

[4] Stats. 1923, sec. 102.35.

[5] Acts of 1915-16, sec. 40, amended 1924, ch. 261.

In other states, the laws grant compensation for diseases listed in schedules, similar to the plan of the British act. The states having schedules of occupational diseases are Minnesota, with 23 diseases listed;[1] New Jersey, 11 diseases;[2] New York, with an almost exact duplicate of the British Act, listing 27 diseases;[3] and Ohio, listing 18 diseases.[4] Porto Rico has a compensation law listing 15 diseases.[5]

The earliest and all-inclusive plan left the word " accidental " out of compensation laws and provided compensation for personal " injuries ". A modified and improved form of the latter has come to be the following provision: " The term ' injury ' as used in this act, shall include any injury or disease arising out of the employment." This permits the granting of compensation for all occupational disease.[6]

The New York Act, as already indicated, schedules all the diseases for which compensation will be granted, as well as the processes in which such occupational diseases may be contracted. In view of the tendency in some states to enact occupational disease schedules, it may be advisable to reproduce the particular section of the New York Act, in order to indicate some of the shortcomings of this method of legislation to meet ever-increasing and heretofore unrecognized occupational diseases.

---

[1] *Acts of 1921*, ch. 82, pt. ii, sec. 67.

[2] *Acts of 1911*, ch. 95, added 1924, ch. 124, sec. 2.

[3] *Consol. Laws*, ch. 67, added by 1914, ch. 41, as amended 1920, ch. 538; 1922, ch. 615; 1928, ch. 754; 1929, ch. 298.

[4] *Genl. Code*, secs. 1465-1468a, added 1921, p. 181, as amended 1929.

[5] *Acts of 1928*, Act no. 85, sec. 3.

[6] *Amer. Labor Legis. Rev.*, vol. 17, 1927, p. 263.

## New York

Compensation shall be payable for disabilities sustained or death incurred by an employee resulting from the following occupational disease:[1]

| *Description of Disease* | *Description of Process* |
| --- | --- |
| Anthrax | Handling of wool, hair, bristles, hides or skins |
| Lead poisoning or its sequelae | Any process involving the use of or direct contact with lead or its preparations or compounds |
| Zinc poisoning or its sequelae | Any process involving the use of or direct contact with zinc or its preparations or compounds or alloys |
| Mercury poisoning or its sequelae | Any process involving the use of or direct contact with mercury or its preparations or compounds |
| Phosphorus poisoning or its sequelae | Any process involving the use of or direct contact with phosphorus or its preparations or compounds. |
| Arsenic poisoning or its sequelae | Any process involving the use of or direct contact with arsenic or its preparations or compounds. |
| Poisoning by wood alcohol | Any process involving the use of wood alcohol or any preparation containing wood alcohol |
| Poisoning by benzol or nitro, hydro, hydroxy, and amido derivatives of benzene (dinitro-benzol, anilin and others), or its sequelae | Any process involving the use of or direct contact with benzol or nitro, hydro, hydroxy or amido derivatives of benzene or its preparations or compounds |
| Poisoning by carbon bisulphide or its sequelae, or any sulphide | Any process involving the use of or direct contact with carbon bisulphide, or its preparations or compounds, or any sulphide |
| Poisoning by nitrous fumes or its sequelae | Any process in which nitrous fumes are evolved |

[1] *Consol. Laws*, ch. 67, added by 1914, ch. 41; as amended 1920, ch. 538; 1922, ch. 615; 1928, ch. 754; 1929, ch. 298; 1930, ch. 60.

| | |
|---|---|
| Poisoning by nickel carbonyl or its sequelae | Any process in which nickel carbonyl is evolved |
| Dope poisoning (poisoning by tetrachlor-methane or any substance used as or in conjunction with a solvent for acetate of cellulose or nitro cellulose), or its sequelae | Any process involving the use of or direct contact with any substance used as or in conjunction with a solvent for acetate of cellulose or nitro cellulose |
| Poisoning by formaldehyde and its preparations | Any process involving the use of or direct contact with formaldehyde and its preparations |
| Chrome ulceration or its sequelae or chrome poisoning | Any process involving the use of or direct contact with chromic acid or bichromate of ammonium, potassium, or sodium or their preparations |
| Epitheliomatous cancer or ulceration of the skin or of the corneal surface of the eye, due to tar, pitch, bitumen, mineral oil or paraffin, or any compound, product, or residue of any of these substances. | Handling or use of tar, pitch, bitumen, mineral oil, or paraffin or any compound, product, or residue of any of these substances |
| Glanders | Care or handling of any equine animal or the carcass of any such animal |
| Compressed-air illness or its sequelae | Any process carried on in compressed air |
| Miners' diseases, including only cellulitis, bursitis, ankylostomiasis, tenosynovitis and nystagmus. | Any process involving mining |
| Cataract in glassworkers | Processes in the manufacture of glass, involving exposure to glare of molten glass |
| Radium poisoning or disability due to radio-active properties of substances or to Roentgen rays (X-rays). | Any process involving the use of or direct contact with radium or radio-active substance or the use of or direct exposure to Roentgen rays (X-rays). |

| | |
|---|---|
| Methyl chloride poisoning | Any process involving the use of or direct contact with methyl chloride or its preparations or compounds |
| Carbon monoxide poisoning | Any process involving direct exposure to carbon monoxide in buildings, sheds or inclosed places |
| Poisoning by sulphuric, hydrochloric or hydrofluoric acid | Any process involving the use of or direct contact with sulphuric, hydrochloric, or hydrofluoric acids or their fumes |
| Respiratory, gastro-intestinal or physiological nerve and eye disorders due to contact with petroleum products and their fumes. | Any process involving the use of or direct contact with petroleum or petroleum products, and their fumes |
| Disability arising from blisters or abrasions | Any process involving continuous friction, rubbing or vibration causing blisters or abrasions |
| Disability arising from bursitis or synovitis | Any process involving continuous rubbing, pressure or vibration of the parts affected |
| Dermatitis (venenata) | Any process involving the use of or direct contact with acids, alkalies, acids or oils capable of causing dermatitis (venenata). |

From the point of view of administration, both methods, the blanket law and the schedule system of compensation, have their definite drawbacks. There is a tendency as indicated by recent legislation, to enlarge the scope of compensation laws by continually increasing the number of compensable diseases. At the same time, the difficulties of medical diagnosis of the occupational diseases, and of estimating the degree of disability, have made the administration of these laws vexatious and not infrequently absurd. This condition of affairs causes unnecessary hardships to both the claimant for compensation and his employer.

The blanket law is theoretically assumed to be the best, and appears to work well where in use. There is, however, a definite opinion among those directly concerned with compensation work that definite and specific schedules facilitate administration, and are beneficial to the industrial worker in that they tend to promote legitimate compensation cases.[1] The schedules may be unfair where the occupation or process is specified, for one individual might secure compensation for a scheduled disease, and another with the same disease refused such compensation because contracted in an unlisted occupation or process.[2]

### INCIDENCE OF OCCUPATIONAL DISEASE COMPENSATION

Some figures are available to indicate the working of the occupational disease compensation laws. The New York law gives compensation for occupational diseases listed in the preceding schedule, though dermatitis, a rather frequent condition, is not recognized as an occupational disease *per se,* unless the substance from which the worker contracts the dermatitis is mentioned in the schedule. A worker suffering from dermatitis, therefore, has an advantage if he contracts the disease from a substance mentioned under the law. If the substance is not mentioned under the law he does not receive compensation although he is just as much disabled, loses just as much time and wages.

[1] The Board of Commissioners for the Promotion of Uniformity of Legislation in the United States, recommended an occupational disease law modeled after the British Act of 1906, with a definite schedule of diseases and processes. *U. S. Bur. Lab. Stat. Mo. Lab. Rev.,* vol. vi, no. 6, June, 1918, pp 219-221.

[2] The Draft Convention Concerning Workmen's Compensation for Occupational Diseases, adopted at Geneva, in 1925, provided for a schedule enumerating the diseases and poisonings, as well as the trades and industries, similar to British act. Natl. Ind. Conf. Bd., *The Work of the International Labour Organization,* 1928, p. 44.

In an analysis of 134 dermatitis cases heard at occupational disease calendars in New York from July 1, 1927 to June 30, 1928, it was found that 88 or 65.7 per cent were disallowed for several reasons, 71 or 53.0 per cent of the total not being covered by the act.   Only 23 or 17.2 per cent received compensation.[1]

It is interesting to note the comparatively few cases that have come up for compensation under the occupational disease laws.   In New York, from July 1, 1927 to June 30, 1928 only 266 cases applied for compensation, and of these, 53 per cent were disallowed for various reasons; 26 per cent were given awards; 18 per cent were still pending at the end of the year, and 3 per cent were disposed of otherwise.   The substances causing the occupational diseases included lead, 65 cases; dye 39; soaps and cleansers 18; acid 15; carbon monoxide 14; volatile hydrocarbons 13; chromic acid 11; miscellaneous 65 cases.[2]

In Ohio during the year 1927 a total of 1,051 occupational disease claims were filed.   Most of the cases, amounting to 696 or 66.2 per cent, were due to dermatitis, many of which resulted in no loss of time.   There were 165 lead poisoning claims and other claims filed covered poisoning by mercury, arsenic, copper, benzol, anilin and phosphorus.[3]

In Wisconsin, for the year 1928, a total of only 395 cases received compensation.   In this state the causative agents of the occupational diseases for which compensation was granted included the following: metallic poisons 37; toxic gases, vapors, fumes, 34; toxic fluids 124; irritant dusts and fibers 62; germs 26; miscellaneous irritants 47; air compres-

[1] *The Industrial Bulletin*, N. Y. State Dept. of Labor, vol. viii, no. 3, Dec., 1928, pp. 486-487.

[2] *Ibid.*, vol. viii, no. 9, June, 1929, pp. 660-661.

[3] *U. S. Bur. Lab. Stat. Mo. Lab. Rev.*, vol. xxvii, no. 2, Aug., 1928, pp. 53-54.

sion 5; miscellaneous 60.[1]   In Massachusetts for 1928, there was a total of 292 cases.   The diseases were: industrial dermatitis 123; lead poisoning 41; gas poisoning 30; acid and oil fumes 22; chrome poisoning 27; cyanide poisoning 5; benzol poisoning 4; anthrax 7; others 33.[2]   In New Jersey, there were only 150 occupational disease cases for which compensation was paid during the year ending December 31, 1928.   Of the total, lead poisoning was the cause in 68 cases and benzol in eight cases.[3]

From the few cases of occupational disease compensation granted in these and other states, the conclusion has erroneously been drawn by somè that the total number of such cases cannot be large, and therefore no serious problem faces the workers in American industry.   Unfortunately, as has already been indicated, many cases of occupational disease are not easily recognizable; or the worker changes his occupation before the occupational disease which he has contracted becomes sufficiently aggravated to seriously incapacitate him; or medical examiners are not capable of determining the cause of disability which may not develop for many years after the early occupation has been changed.

The development in the United States of laws granting compensation for occupational diseases, whether through so-called blanket legislation, or by definite schedules of occupational diseases, represents, in the last analysis, a wholesome though only an initial attempt to meet the problem of hazards to the health and life of the worker.   Some shortcomings of these methods have been indicated.   The failure to

---

[1] Industrial Commission of Wisconsin, *Wisconsin Labor Stat. Bull. no. 20,* Sept. 5, 1929, p. 11.

[2] Massachusetts Dept. of Labor and Industry—Division Industrial Safety, *Report for Year ending Nov. 30, 1928* (Boston), pp. 22-23.

[3] New Jersey Dept. of Labor, *The Industrial Bull.* (Trenton, June, 1929), p. 7.

include occupational diseases in early American, and to some extent, in present compensation acts, has probably been due, in part at least, to the lack of definite information as to their prevalence. Unfortunately, many workers are sacrificed under the present system of waiting until an occupational disease becomes fairly widespread before an attempt is made to provide a remedy and to grant compensation to the injured workers or their surviving dependents.

Further, the discrimination indicated in schedules of occupational diseases and industrial processes, causes additional aggravation of an unsatisfactory state of affairs. It may not be too much to assume that some day the constitutionality of legislation which permits compensation, for instance, for carbon bisulphide poisoning and not for hydrogen sulphide poisoning, will be seriously questioned. More careful analysis and study of the shortcomings of both methods of occupational disease compensation, which at present involve a lack of definite standards in both medical and technical aspects, may lead to additional or other methods of safeguarding the health and interests of the industrial workers.

### JUDICIAL DECISIONS AFFECTING OCCUPATIONAL DISEASES

From the beginning of workmen's compensation administration, a broad distinction has been made between accidental injury and occupational disease. The accidental injury is considered an unforeseen, unexpected event, occurring suddenly, and in some jurisdictions requiring violence to the physical structure of the body; whereas an occupational disease is reasonably to be expected as a natural result of the hazards of the employment and develops gradually. This distinction is emphasized in an Illinois case, in which the court said: [1]

[1] Matthiessen and Hegeler Zinc Co. *v.* Industrial Board, 284 Ill. 378, 120 N. E. 249 (1918).

A disability caused in that way or from that source (occupational disease) is not to be regarded as an accident because such a disease has its inception in the occupation and develops over a long period of time from the nature of the occupation and not from any unusual or unforeseen cause or event.

The English Compensation Act which served as a model for several of the compensation laws in this country originally excluded compensation for occupational diseases. Following this example, the early compensation acts enacted by various states in this country, generally restricted compensation to accidental injuries. That the usual course of legislation has been to deal with industrial accidents as separate and distinct from occupational diseases is explained in a Michigan case.[1] It was held here that because the paramount object of the enactment of the compensation laws was to afford a more just and humane remedy for accidental injuries than was afforded by the common law, by removing certain defenses, and inasmuch as under the common law an employee had no right of action for injury or death due to occupational disease, it would seem that occupational diseases were not intended to be included in the compensation acts.

The same idea is expressed in a Connecticut case,[2] in which the court held that:

Since the common law action for damages which was founded on the master's negligence never attempted to cover the typical case of an occupational disease caused by continued exposure to the ordinary and known risks of the employment, the inference is plain that the alternative compensation scheme was not intended to cover such diseases. . . . It may be said that in point of logic occupational disease is as proper a subject for compensation as industrial accident.

[1] Adams *v.* Acme White Lead Co., 182 Mich. 157, 148 N. W. 485 (1914).
[2] Miller *v.* American Steel and Wire Co., 90 Conn. 349, 97 Atl. 345 (1916).

### EMPLOYER'S LIABILITY

Liability under the American workmen's compensation laws is created by a standardized phrase, taken bodily from the English act and incorporated in the majority of our state laws—" personal injury by accident arising out of and in the course of employment." The courts, in attempting to define this liability, have varied widely in their opinions, depending, largely upon whether a strict and literal, or broad and liberal interpretation of the law has been made. In some states, moreover, a variation in the phrasing of this clause results in vital differences in interpretation. Where the word " injury " is used alone it has generally been held to cover a broader scope than where the element of accident must accompany the injury in order that it may be compensable.[1] " Injury " where used alone has been held to include all accidents, but the word " accident " does not include all injuries.[2]

In order that a disability be termed " personal injury " it is not necessary that there be a blow or sudden physical contact with the body, producing violence thereto. The Massachusetts Supreme Judicial Court, in sustaining an award to a workman blinded from inhaling poisonous gas held that: " The preponderance in recent years of actions grounded upon physical violence has tended to emphasize the aspect of injury which depends upon visual contact or direct lesion. But this is by no means the exclusive significance of the word either in common speech or in its legal use." [3]

Quite naturally, there have been many and various decisions dealing with this subject. An interesting definition of

---

[1] Johnson *v.* London Guarantee and Accident Co., 217 Mass. 388, 104 N. E. 735 (1914).

[2] Cooke *v.* Holland Furnace Co., 200 Mich. 192, 166 N. W. 1013 (1918).

[3] *In re* Hurle, 217 Mass. 223, 104 N. E. 336 (1914).

the matter was rendered in a Massachusetts case,[1] in which the court held that:

Anything which causes incapacity for work and thereby impairs the ability to earn wages is a personal injury under the Workmen's Compensation Act. Generally classed as personal injuries are all the consequences of an accident, which may also include resulting infection or diseases, like traumatic pneumonia, or an aggravation or acceleration of a pre-existing disease.

It is seemingly apparent from these few citations that there is not, as there necessarily cannot be, any unanimity of opinion regarding the liability of the employer, and the interpretation of the terms " injury " and " accident ". The situation is still further complicated by the lack of uniformity in legislation in the various states, for, as already noted, many have no provision for compensation of occupational diseases, some provide for such compensation by so-called " blanket " legislation, and a few others by schedules of occupational diseases.

### AGGRAVATION OF PRE-EXISTING CONDITIONS

The pre-existence of disease in compensation cases gives rise to problems for which it is exceedingly difficult to find a solution which will be equitable to all concerned. In general, the accepted theory seems to be that an employer takes workers as he finds them and must, therefore, assume the burden of compensating an injured employee even though disease or disability, unrelated to the employment, increases the probability of serious consequences in case of accidental injury. To some extent employers, particularly in large scale establishments, have fortified themselves by requiring applicants for employment to undergo a physical examination prior to employment, by establishing an age limit, and,

[1] *In re* Johnson, 217 Mass. 388, 104 N. E. 735 (1914).

in some instances, requiring a periodic medical re-examination.[1]

The courts have held with considerable uniformity that where an employee afflicted with disease receives a personal injury for which he might be entitled to compensation had there been no disease involved, he may be awarded compensation if, as a result of the injury, the disease is accelerated or aggravated, and its progress materially contributing to hasten its culmination in disability or death.[2]

Mere disposition to disease does not make the injury any the less accidental, if the disease is actually brought on by the injury.[3] Where a fireman fell from an engine cab and died without regaining consciousness, although the autopsy showed pre-existing disease which predisposed him to hemorrhage of the brain, which caused his death, the court held that the accident was a contributing cause, and an award for compensation was sustained.[4] The mere fact that the employee's condition renders him more susceptible to the particular injury which he sustains, has been held not to be ground for maintaining that the disease or condition, rather than the accident, was the proximate cause of the injury.[5]

It has also been held that if an accident arouses latent germs of a disease to which the workman is predisposed, materially accelerating the disease and causing death, it is an accident within the law.[6] Further, the presence of a disease which in itself partially disables the worker, has been held

[1] National Industrial Conference Board, *The Workmen's Compensation Problem in New York*, 1927, p. 80.

[2] *In re* Bowers, Williams, Colan, 65 Ind. App. 128, 116 N. E. 842 (1917).

[3] Wabash R. Co. *v.* Industrial Commission, 286 Ill. 94, 121 N. E. 569 (1918).

[4] Peoria R. Co. *v.* Industrial Board, 279 Ill. 352, 116 N. E. 651 (1917).

[5] Puritan Bed Spring Co. *v.* Wolfe, 68 Ind. App. 330, 120 N. E. 417 (1918).

[6] Retmier *v.* Cruse, 67 Ind. App. 192, 119 N. E. 32 (1918).

not to operate to deprive him of compensation where it is shown that the accident increased the disability.[1]  It is sufficient to justify an award if the accident, by weakening resistance, or otherwise influences existing disease, causes disability or death.[2]  It has been held, however, that a slight acceleration or aggravation of an existing, progressive disease, is not sufficient to bring the case within the spirit or letter of the compensation law.[3]

In a case where an employee sustained an injury to his head, and due to a pre-existing condition of syphilis, he subsequently became insane, it was held that compensation should be granted.[4]  Where a pre-existing heart disease of a servant was accelerated to the point of disablement by the muscular exertion required by the employment, the court found that she suffered a personal injury in the course of employment and was entitled to compensation.[5]

Many of the cases which have reached the courts in the matter of aggravation or acceleration of pre-existing disease, have usually arisen out of an accidental injury, rather than because of occupational diseases.  This may be due largely to the fact that only some of the states compensate for occupational diseases, and also to the circumstance that occupational diseases too frequently go unrecognized as such and therefore few claims for aggravation or acceleration as a result of occupational disease are apt to arise.  With more adequate methods of discovering and diagnosing occupational diseases in their incipiency, more cases will probably

[1] Slinger *v.* Muskegon Motor Specialties Co., 201 Mich. 473, 167 N. W. 949 (1918).

[2] Mailman *v.* Record Co., 118 Me. 172, 160 Atl. 606 (1919).

[3] Borgsted *v.* Shultz Bread Co., 167 N. Y. Supp. 647, 180 App. Div. 229 (1917).

[4] In re Crowley, 223 Mass. 288, 111 N. E. 786 (1916).

[5] Madden's Case, 222 Mass. 487, 111 N. E. 379 (1916).

reach the courts. From present indications, it seems likely that the courts will tend to give liberal interpretations to the laws and generally to protect the interests of the workers.

### JUDICIAL DECISIONS

Some indication has already been given of the difficulty in differentiating between occupational diseases and injuries, as affecting claims for workmen's compensation or for damages under common law actions. In the following cases the attempt is made to avoid, as far as possible, such cases as are definitely the result of accidental injuries and which cannot be classified in any sense as arising out of an occupational disease. However, the distinction cannot be drawn too finely, for in certain instances, neither legal nor medical opinion is unanimous as to where the line of demarcation is to be placed.

### ANTHRAX

In an English case decided in 1905 where a workman contracted anthrax by a germ settling on his eye while sorting wool which was infected with anthrax, it was held that he had suffered injury by accident and was entitled to compensation. The court further held that: "It does not appear that by calling the consequence of an accidental injury a disease, alters the nature or the consequential results of the injury that has been inflicted." [1] It should be noted that this case was decided before anthrax was placed on the British list of diseases for which compensation was to be granted. Subsequent decisions in the United States have in a number of cases followed the precedent set and reasoning in this case.

In a New York case an employee contracted anthrax from handling hides, the germs entering a fissure on his hand, caused by a previous handling of hides. The disability

[1] Brinton's Ltd. *v.* Turvey, A. C. 230, 7 W. C. C. I. (1905).

caused by the anthrax was held to be due to an accident and compensation was granted.[1] It was held to be an error to grant compensation to an employee who received a cut on his neck while not engaged in his duties, and died from anthrax while employed in a tannery.[2] In a similar case of a tannery employee, a piece of tissue cut from the swollen pimple on his neck was found to be infected with anthrax bacilli which evidently caused his death. The court held that the acquiring of the anthrax bacillus from the hides in the tannery was the only reasonable explanation of the infection; that the disease was an accident arising out of the employment and an award for compensation was affirmed.[3] The difference of judicial opinion in these tannery cases is indicative of some of the problems arising in the adjudication of occupational disease cases, whether they be looked upon as accidental injuries or as occupational diseases.[4]

[1] Heiers *v.* Hull and Co., 164 N. Y. Supp. 767, 178 App. Div. 350 (1917).

[2] Eldridge *v.* Endicott, Johnson and Co., 189 N. Y. App. Div. 53, 126 N. E. 254 (1920).

[3] Chicago Rawhide Mfg. Co. *v.* Industrial Commission, 291 Ill. 616, 126 N. E. 616 (1920).

[4] Houston Packing Co. *v.* Mason, Tex. Civ. App., 286 S. W. 862 (1926). This was another anthrax case in which the court held as follows: " The authorities are not in harmony as to whether death resulting from such disease contracted by an employee in the course of his employment is compensable as for an injury arising by accident. The great weight of authority holds that in such cases compensation should be allowed, and the authorities so holding are sounder in principle and more in consonance with the intent and reason of the law and the liberal interpretation to be given it than those holding to the contrary. In such cases it is because the employee in the discharge of his duty is brought in contact with the anthrax germ that he contracts the disease, and it is correct to say that the consequent injury suffered by the employee was accidental and the result of a hazard incident to and in the course of his employment and therefore compensable."

## APOPLEXY

This is a condition about which the courts and industrial boards frequently disagree as to whether or not it falls among the so-called occupational diseases. An employee, while assisting another to lift a 200-pound barrel, was seized with a stroke of apoplexy, by reason of the unusual strain occasioned by the lifting. This was held by the New York courts to be a compensable accident.[1] Where there was no accidental injury which aggravated or brought on the apoplexy, it was held not to be accidental, and therefore compensation was denied.[2] A teamster accidentally fell from his wagon, receiving slight injuries. Three hours later he suffered a stroke, resulting in paralysis. Though most of the medical experts testified there was no connection between the fall and the stroke, compensation was allowed.[3]

A laborer died suddenly in bed five or six months after an accident that had seriously injured his head and body. Medical opinion held that apoplexy may have caused the death. The court affirmed an award of compensation.[4]

In cases in which pre-existing arterioslerosis has been the main cause of disability or death, the courts have held that where a cerebral hemorrhage occurred, and was the result of arteriosclerosis, and may have been superinduced by heat prostration, a preponderance of evidence to prove the casual relationship must be presented or compensation cannot be granted;[5] physical exertion in the course of employ-

[1] Fowler *v.* Risedorph Bottling *et al.,* 175 N. Y. App. Div. 224, 161 N. Y. Supp. 535 (1916).

[2] Ledoux *v.* Employer's Liability Assur. Corp., 2 Mass. Ind. Acc. Board 493.

[3] Selaya *v.* Ruthuen and Cerrano, 5 Cal. Ind. Acc. Comm. 238.

[4] Judice *v.* Degnon Cons. Co., 167 N. Y. Supp. 1107, 181 N. Y. App. Div. 909 (1917).

[5] Stunnette *v.* Gillespie Co., 2nd Rep. Ky. Leading Dec., p. 5.

ment producing an increase in blood-pressure resulting in death of an employee afflicted with arteriosclerosis, was subject to compensation;[1] where it did not appear that a cerebral hemorrhage was due to an unusual strain in the course of employment, but rather to pre-existing arteriosclerosis, compensation was denied;[2] a severe injury to an ankle of a worker during employment, with subsequent death from apoplexy, was held to be a contributing cause and compensation was granted.[3]

### CAISSON DISEASE

Mention has already been made of the states in which caisson disease or " the bends " is compensated for under the occupational disease provisions of the workmen's compensation acts. In a case where the employee charged with regulating the reduction of air pressure in the lock negligently reduced it too rapidly, with the result that a worker sustained fatal injuries, the appeal from an award for compensation on the ground that the injury was caused by conditions peculiar to the industry in which the decedent worker was engaged, and hence due to an occupational disease, which in this instance was not compensable, was denied. The court held that the death was due to an accident, and that the case was not to be classed among the occupational diseases, which ordinarily require considerable time for development.[4]

[1] Samoskie *v.* Phila. and Reading Coal and Iron Co., 280 Pa. 203, 124 Atl. 471 (1924).

[2] *In re* Mrs. Alfred Haries, vol. 1, no. 7, Bull. Ohio Ind. Comm. 101.

[3] U. S. Casualty Co. *v.* Matthews, 35 Ga. App. 526, 133 S. E. 875 (1926).

[4] Williams *v.* Missouri Valley Bridge and Iron Co., 212 Mich. 150, 180 N. W. 357 (1920).

## CANCER

Cases of cancer arising out of accidental or other injuries received in the course of employment, have presented a number of difficult problems to the courts, including the conflict of medical testimony.   A worker presented a claim for disability due to cancer of his left clavicle which he contended was the result of a fall.   The industrial commission granted compensation.   On appeal, there was conflicting medical testimony, the court holding however, that as the testimony was conflicting, the affirmative award would not be disturbed.[1]   The dependents of a worker who developed cancer of the lip after being struck by a hot piece of iron, were denied compensation, after it had been proved that prior to his death from cancer he had a sore on his lip, the court stating: " The evidence simply shows that the sore on the lip developed into cancer that caused the death, and there was insufficient proof that the burn was an accelerating or proximate cause of the death." [2]

In a number of cases accidental injuries were held to have aggravated pre-existing cancerous conditions.[3]   In instances where no direct relationship between the injury and the subsequent cancer could be satisfactorily established,[4] or where it was uncertain whether death was due to an injury or pre-existing disease, compensation has been denied.[5]

[1] Sugar Co. of Santa Ana *v.* Industrial Acc. Comm., 35 Cal. App. 652, 170 Pac. 630 (1918).

[2] West Side Coal and Mining Co. *v.* Industrial Comm., 321 Ill. 61, 151 N. E. 593 (1926).

[3] Broussard *v.* Union Sulphur Co., 5 La. App. 340, La. Digest 160 (1927) ; Voorhees *v.* Smith, Schoomaker Co., 86 N. J. 500, 92 Atl. 280 (1914) ; Whittle *v.* National Aniline and Chemical Co., 266 Pa. 359, 109 Atl. 847 (1920).

[4] McElligott *v.* Frankfort General Ins. Co., 2 Mass. Ind. Acc. Bd. 521.

[5] Malvern Lumber Co. *v.* Sweeney, 116 Ark. 561, 172 S. W. 821 (1914).

### DERMATITIS

Statistics of compensation boards indicate that dermatitis is at present the cause of many of the claims for compensation. It is frequently difficult to draw the line between accidental injury and occupational disease as a cause of dermatitis. Where a chambermaid contracted dermatitis of the hands, it was held that the resulting disability was not compensable, because there was no proof of its accidental origin.[1] A worker developed dermatitis of the lower limbs, and it was held that the infection was caused by walking back and forth all day in wood dust to the depth of an inch. This was held to be a personal injury.[2]

In a New York case,[3] a worker was employed to handle and brush dry furs after the process of dyeing was complete. He contracted dermatitis, and claimed compensation due to aniline poisoning.[4] On appeal from an award by the State Industrial Board and in reversing the award the court held:

Any process involving the use of aniline is descriptive of a chemical process, and in the fur dyeing trade involves the application of the chemical to the furs. The claimant, not having been engaged in making the application of the dyes, was not within the coverage of the law and was not suffering from an occupational disease.

[1] McDonald *v.* Dunn, 2 Cal. Ind. Acc. Comm. Dec. 91.

[2] Reeves *v.* Diamond Match Co. *et al.*, 5 Cal. Ind. Acc. Comm. Dec. 236.

[3] Sokol *v.* Stein Fur Dyeing Co. *et al.*, 216 N. Y. Supp. 167, 216 App. Div. 573 (1926).

[4] See *Consol. Laws of New York*, 1922, ch. 615. Ch. 754, effective July 1, 1928, made occupational disease compensable when due to "direct contact with" the several poisonous substances enumerated in the compensation law. The amended phraseology reads: "use of or direct contact with", the words "or direct contact with" having been added. This was a direct outgrowth of decisions in cases similar to Sokol *v.* Stein Fur Dyeing Co. The amendment of Ch. 754 also extended the contracting of disease to continuous employment similar to the one in which the employee incurs disablement.

In a similar case the court held that the particular work did not relate to a " process involving the use of " dyes and chemical, as specified in the law, and denied compensation.[1]

Evidence was required in order to show that an employee's dermatitis was occupational in nature and arose from cement poisoning.[2] Disability due to dermatitis, caused by contact with sulphuric acid, without accidental injury, was held not to be an " occupational disease ", and therefore not compensable.[3]

Where an employee after ten days of service in a bleachery developed a skin condition, it was held not to be an accidental injury and therefore not compensable.[4]

The scheduling of occupational diseases in compensation laws, with the stipulation " process involving the use of " certain chemicals, prevents the granting of seemingly just claims for compensation.

DUST

Several types of cases have arisen in which some kind of dust has been the cause of disability.   In a Tennessee case,[5] it was held that under the workmen's compensation act of the state, a disease caused by breathing dust as a result of moving sacks containing a chemical used in the employer's business is not a compensable injury.   A disease caused by breathing fine metallic and mineral particles thrown into the air where an employee was working, was held not to be the result of an accident, and that recovery for injuries therefore is not confined to the workmen's compensation act.

[1] Amsterdam *v.* Hammer Bros., 210 App. Div. N. Y. 816 (1924).

[2] Kosick *v.* Manchester Const. Co., 106 Conn. 107, 136 Atl. 870 (1927).

[3] Wright *v.* Used Car Exchange, 223 N. Y. S. 245, 221 App. Div. 154 (1927).

[4] Liondale Bleach, Dye and Paint Works *v.* Riker, 85 N. J. 426, 89 Atl. 929 (1914).

[5] Meade Fiber Corp. *v.* Starnes, 147 Tenn. 362, 247 S. W. 989 (1923).

Where an injury does not fall within the workmen's compensation act, the common law remedy is not affected by it.[1] Where the disability of an employee arose from continued breathing of iron dust in his particular occupation, and there was no one circumstance, incident or time to which could be pointed the starting place of such disability " it cannot be held to arise out of an accidental injury within the compensation act, but was a disease of occupation and not compensable." [2]

Dust in industry usually effects the respiratory tract, and therefore there have been a number of cases involving tuberculosis and other lung conditions. These cases will be considered under a subsequent heading.

### EYE DISEASES

An operator of sheet metal finishing rolls was blinded and permanently disabled by the strong glare of powerful lights from the glittering surfaces he had to inspect; it was held he was suffering from an occupational disease and was not entitled to compensation.[3] Injuries to eyes have been compensated for, even though the injured eye was defective before the accidental injury, on the ground that the law does not specify a normal eye as the basis of compensation awards.[4]

Under the Massachusetts Workmen's Compensation Act providing for compensation to an employee receiving a " personal injury arising out of and in the course of his em-

[1] Smith *v.* International High Speed Tool Co., 98 N. J. 574, 120 Atl. 188 (1923).

[2] Peru Plow and Wheel Co. *v.* Industrial Comm., 311 Ill. 216, 142 N. E. 546 (1924).

[3] Zajkowski *v.* American Steel and Wire Co., 258 Fed. 9 (1918).

[4] Purchase *v.* Grand Rapids Refrigerator Co., 194 Mich. 103, 160 N. W. 391 (1916).

ployment," it was held that an employee suffering from total loss of vision in both eyes, resulting from an acute attack of optic neuritis induced by poisonous coal gases escaping from furnaces about which he worked, was entitled to compensation.[1] A partial and temporary loss of sight, caused by the bursting of small blood vessels in the eye, due to increased blood pressure resulting from heavy lifting was held to be compensable.[2] In a somewhat similar case, in which blindness followed exposure to intense bright light, it was held that the blindness was due to natural causes, and not to an accident arising out of the employment.[3] Under the federal workmen's compensation act conjunctivitis caused by working in a very strong light, was held to be a compensable injury in one case,[4] and not compensable in another similar case.[5]

### GLANDERS

A stableman was required to lead a horse afflicted with glanders from his employer's stable to a place where it was to be killed. Fourteen days after performing this service he died from glanders. An award of compensation was made upon findings that the employee had contracted glanders, and that his infection was an accidental injury arising out of and in the course of his employment. On appeal the second point was questioned. The court held: "For legal purposes glanders is a disease which, when contracted without previous accidental injury occurring in the course of employment, cannot be classed under the workmen's compensation law as an accidental injury arising out of and in the

[1] *In re* Hurle, 217 Mass. 223, 104 N. E. 336 (1914).
[2] Gurney *v.* Los Angeles Soap Co., 1 Cal. Ind. Acc. Comm. Dec. 163.
[3] Crouch *v.* Ritter, 2 Cal. Ind. Acc. Comm. Dec. 702.
[4] *In re* H. E. Cuttright, 3rd A. R. U. S. C. C. 119.
[5] *In re* James Hawthorne, 3rd A. R. U. S. C. C. 120.

course of employment." The award of compensation was reversed.[1]

### HEART DISEASE

Cases of heart disease are of frequent occurrence among those of working age, and therefore, as might be expected, of similar frequency among the claims for compensation. Some mention has already been made of the action of the courts in cases of pre-existing disease aggravated by the nature of the work. Whether or not such heart conditions may be classified in certain instances as occupational diseases is often difficult to determine. It has been held that the breaking down of heart muscles from strain in employment is compensable, although the employee's physical condition predisposed him to heart disease;[2] an injury resulting from a fall in the course of employment may be a contributory cause of final breakdown and is compensable;[3] death from heart disease in the course of employment caused by over-exertion due to the employment, so as to be an "accident" is compensable;[4] an employee to be entitled to compensation for injury, must have been exposed by employment to risks more than normal to the general public. Death of a miner, predisposed to heart disease, while passing through an air shaft on his way from work was held to be an accident;[5] direct casual connection between personal injury to a worker

[1] Richardson *v.* Greenberg, 188 N. Y. App. Div. 248, 176 N. Y. Supp. 651 (1919).

[2] Mellquist *v.* Dakota Printing Co., 51 S. D. 359, 213 N. W. 947 (1927).

[3] Ocean Accident and Guar. Corp. *v.* Industrial Comm. of Utah, 66 Utah 600, 245 Pac. 343 (1926).

[4] Ellerman *v.* Industrial Comm. of Colo., 73 Colo. 20, 213 Pac. 120 (1923) ; Southern Casualty Co. *v.* Flores. Texas Ct. Civil Appeals, March, 1927, 294 S. W. 932.

[5] Nicholson *v.* Roundup Coal Mining Co., Sup. Ct. Mont., June, 1927, 257 Pac. 270.

and the employment must be established to entitle the worker to compensation; [1] the same state court held in 1918 than an injury need not be an accident in order to be compensable, as where a pre-existing heart disease was accelerated to the point of disablement due to the exertion required by the employment; [2] death from heart disease through rupture of a diseased aorta was a personal injury resulting from an accident. [3]  In another case involving death due to rupture of the aorta the Pennsylvania Court held that injury need not rise out of or be due to the worker's employment, as it is sufficient if it happens in the course thereof. [4]  The death of an employee from heart disease precipitated by high altitude and over-exertion, even though it is proved that the employee had a weak heart, is compensable; [5] enlargement and dilation of the heart, resulting from employment in excessive heat, was due to an accident; [6] where an employee had organic heart trouble, and the strenuous work of pitching alfalfa hay in an enclosed building, combined with breathing dust-laden air, brought on an attack of heart trouble, causing death, the condition of the air or the fact that it was dust-laden was held to be the proximate cause of the death. [7]

[1] Madden's Case, 222 Mass. 487, 111 N. E. 379 (1916).

[2] *In re* Mooradjian, 229 Mass. 521, 118 N. E. 951 (1918).

[3] Indian Creek Coal and Mining Co. *v.* Calvert, 68 Ind. App. 474, 119 N. E. 519 (1918).

[4] Clark *v.* Lehigh Valley Coal Co., 264 Pa. 529, 107 Atl. 858 (1919).

[5] Knock *v.* Industrial Accident Comm. of California, Sup. Ct. Cal., Feb., 1927, 253 Pac. 712.

[6] Becton *v.* Deas Paving Co., M. and S. La. Digest 154, 3 La. App. 683 (1926).

[7] Carroll *v.* Ind. Comm. of Colo., 69 Colo., 473, 195 Pac. 1097 (1921); In McMurray *v.* Little and Ives Co., 3 N. Y. St. Dep. Rep. 395 (1915), an electrotype finisher who was found dead in a chair in a saloon, where he had gone after the day's work, and medical testimony established that he had suffered from angina pectoris caused by overwork, com-

In other cases similar to the preceding ones courts have denied compensation. Where a workman died as a result of strenuous labor which caused the rupture of the aorta, it was held that award under the Workmen's Compensation Act of Michigan on the theory that death was an accidental cause could not be sustained; the workman was engaged in his regular employment and the heart defect had been of long standing.[1] A horseshoer died from angina pectoris after exertion on his job, and the court held the heart failure was not a traumatic accident within the meaning of the workmen's compensation act.[2] Similar decisions have been rendered in cases where strenuous work and long hours were combined with heavy lifting, resulting in death from heart disease,[3] as where a blacksmith, disabled by heart disease after many years of employment for the same firm, had his employment changed to lighter work, finally becoming incapacitated from all labor. The court held that at no time was there any injury in the sense of a definite occurrence which could be located in point of time or place. Any gradual or imperceptible effect on the heart, resulting in final disability could not be compensated.[4] Where an employee was found dead near a bailing press, and there was no evidence proving an accident or accidental injury, the claim being that the heavy work hastened the death by heart disease, there could be no

pensation was to be granted, as the court held that death was due to an accidental injury occasioned by the employment. It is interesting to note that compensation was granted even though death occurred away from the workshop.

[1] Stombaugh *v.* Peerless Wire Fence Co., 198 Mich. 445, 164 N. W. 537 (1917).

[2] Wallins Creek Collieries Co. *v.* Williams, 211 Ky. 200, 277 S. W. 234 (1925).

[3] Guthrie *v.* Detroit Shipbuilding Co., 200 Mich. 355, 167 N. W. 37 (1918).

[4] Longobardi *v.* Sargent and Co., 100 Conn. 383, 124 Atl. 13 (1924).

recovery.[1]  Death from heart disease resulting from fright or worry during the course of employment, unaccompanied by any immediate physical injury was held not to be compensable.[2]

### NEUROSIS

A cigar maker who suffered from a neurosis as a result of his sitting posture while rolling cigars for over 25 years, was denied compensation.  No necessary connection was shown to exist between the work and the posture, so that the induced neurosis could not be regarded as an injury arising out of the employment.[3]  In a similar case the court likewise denied compensation, maintaining that the evidence adduced was insufficient to sustain the finding that pain to which the employee was subjected had been a reasonably necessary result of employment, so that it was an injury peculiar to it.  The reasonable inference was, according to the decision, that the neuralgic pain was due to faulty posture equally liable to arise in any or no employment.[4]

### PNEUMONIA

Many cases have arisen in which compensation was claimed for illness or death due to pneumonia following an accident or exposure.  In such cases of pneumonia as followed accidental injuries the courts have generally allowed compensation, on the basis that an accidental injury was sus-

---

[1] Jakub v. Industrial Comm. et al., 288 Ill. 87, 123 N. E. 263 (1919).

[2] Visser v. Michigan Cabinet Co., Mich. Ind. Acc. Bd. Dec. (No. 3, 1913), 24, 10 N. C. C. A. 1042; O'Connell v. Adirondack Elec. Power Corp., 185 N. Y. S. 455, 193 App. Div. 582 (1920).  For additional heart disease cases in which compensation was denied see: Johnson v. Mary Charlotte Min. Co., 199 Mich. 218, 165 N. W. 650 (1917); Amesbury v. Vacuum Oil Co., 9 N. Y. St. Dep. Rep. 399 (1916); Tucillo v. Ward Baking Co., 180 N. Y. App. Div. 302, 167 N. Y. Supp. 666 (1917).

[3] In re Maggelet, 228 Mass. 57, 116 N. E. 972 (1917).

[4] Pimental's Case, 235 Mass. 598, 127 N. E. 424 (1920).

tained,[1] provided the injury arose out of and in the course of employment.[2]   In the case of overheating with exposure, resulting in subsequent pneumonia and death, it has been held that a personal injury was sustained, on the ground that: " If an injury arises out of and in the course of the employment, it shall be no bar to a claim for compensation that it cannot be traced to a definite occurrence which can be located in point of time and place." [3]   In a somewhat similar case in another jurisdiction, compensation was denied, on the ground that no personal injury was sustained.[4]   Successive wettings of an employee by rain, resulting in his contracting a cold and pneumonia, has been held not to be an injury, not being " damage or harm to the physical structure of the body from disease or infection." [5]   Death from pneumonia contracted by a food packer in a storage warehouse while working in an ice box, was held not the result of an accident, for which compensation could be awarded; the hazard, if any, was a continuous one, and fully known to the employee.[6]   In a Connecticut case, the decedent was compelled to labor in damp and poorly heated buildings, and in rainy weather to ride in an open truck to his work.   By reason of these conditions his resistance was lowered and he

---

[1] Curtis-Warner Corp. *v.* Gorman, Sup. Ct. N. Y., Oct., 1925, 130 Atl. 538; Vogeley *v.* Detroit Lumber Co., 196 Mich. 516, 162 N. W. 975 (1917).

[2] Where the injury is held not to have arisen in the course of employment, compensation has been denied.  Anderson *v.* Baxter, Sup. St. Pa. Feb., 1926, 132 Atl. 358.

[3] Dupre *v.* Atlantic Refining Co., 98 Conn. 646, 120 Atl. 288 (1923).

[4] Hoag *v.* Kansas Independent Laundry, 113 Kan. 513, 215 Pac. 295 (1923).

[5] Texas Employers' Insurance Assn. *v.* Jackson, Tex. Comm. of Appeals, Sec. B, Nov., 1924, 265 S. W. 1027.

[6] D'Oliveri *v.* Austin Nichols and Co. *et al.*, 207 N. Y. S. 699, 211 App. Div. 295 (1925).  See also: Newkirk *v.* Golden Cycle Min. and Red. Co., 79 Colo. 298, 244 Pac. 1019 (1926).

contracted pneumonia from which he died.  The court held that a compensable personal injury need not be traced to a definite happening or event, and it may be caused by accident or disease and includes diseases peculiar to the occupation, except those of a contagious, communicable or mental nature.[1]

### POISONING

Cases of poisoning by various substances are somewhat easily handled in certain of the states where compensation is provided for, when such poisoning is proved to have arisen out of the handling of particular substances.  In states making no provision for such compensation, unless it can be shown that the poisoning was a personal injury and not an occupational disease, compensation is usually withheld.

The poisoning of fingers from dipping the hands in a solution in the development of photographic plates in the course of employment, which dipping occurred some 500 times each day during the week, was held not to be an " accident " for which compensation could be awarded.[2]  In an Illinois case, it was held that the death of a worker from arsenical poisoning in the course of his employment, although resulting from many years' exposure, arose from an " accident " or " accidental injury " under the workmen's compensation act for which the employer was liable.[3]

Poisoning by mine gas has been held compensable in a number of cases, it being held that death was due to an accidental injury.[4]  Carbon monoxide poisoning of garage

[1] De La Pena v. Jackson Stone Co., 103 Conn. 93, 130 Atl. 89 (1925).

[2] Jeffreyes v. Charles H. Sager Co., 191 N. Y. S. 354, 198 App. Div. 446 (1921).

[3] Matthiessen and Hegeler Zinc Co. v. Industrial Board, 284 Ill 378, 120 N. E. 249 (1918).

[4] New River Coal Co. v. Files, 215 Ala. 64, 109 So. 360 (1926); New Marissa Coal Co. v. Industrial Comm., 326 Ill. 116, 157 N. E. 32 (1927); Tarr v. Hecla Coal and Coke Co., 265 Pa. 519, 109 Atl. 224 (1921).

employees has likewise been compensated for on the ground that the poisoning was an accidental injury arising out of employment.[1]

Where an employee of a gas company was engaged in making a service connection and was suddenly overcome by gas, resulting in death, compensation was granted, though not for the disease but rather for the accident. It was held that where the result of an injury is attributable in whole or in part to an accident, the fact that but for the accident the disease of which claimant died would be classed as occupational, will not prevent compensation.[2] Where death was caused by an accidental inhalation of gas, the question whether the health of the worker had been impaired by previous inhalations was held to be immaterial and compensation was allowed.[3]

In an action by an employee against his employer to recover damages for injury to his health caused by sulphuric acid poisoning while working in the bleach room of a tannery, the court held that at common law it was the duty of the employer to know whether fumes arising from a vat in the bleach room are poisonous to an employee; judgment was affirmed for the plaintiff.[4] In a somewhat similar case,

[1] Cantor *v.* Elsmere Garage *et al.*, 212 N. Y. S. 327, 214 App. Div. 351 (1925) ; Columbine Laundry Co. *v.* Industrial Comm. of Colo., 73 Colo. 397, 215 Pac. 870 (1923).

[2] Van Vleet *v.* Public Service Co. of York, 111 Neb. 51, 195 N. W. 467 (1923). For a similar case occurring at a gas plant, and for which compensation was granted on the basis of an injury suffered, see: Holnagle *v.* Lansing Fuel and Gas Co., 200 Mich. 132, 166 N. W. 843 (1918).

[3] U. S. Fidelity and Guaranty Co. *v.* Industrial Comm. of Colo., 76 Colo. 268, 230 Pac. 624 (1924). See also: Tintic Milling Co. *v.* Industrial Comm. of Utah, 60 Utah 14, 206 Pac. 278 (1922), in which it was held that disability of disease, either caused or accelerated by an occurrence in the course of employment which is an accident (gas poisoning), is compensable.

[4] Fritz *v.* Elk Tanning Co., 258 Pa. 180, 101 Atl. 958 (1917).

the employee failed to recover. The claimant's employer was a dye manufacturer, and in the shop there were fumes of sulphuric oxide. On a given day, according to the claimant, he received an accidental injury in that he inhaled fumes that were stronger than usual, and that by reason thereof he was totally disabled for some time. He had worked for six years in the same employment and throughout that time he had felt the effects of the fumes. The court held that the particular experience did not cause the disease (chronic bronchitis) complained of, and that no time or place could be fixed when the disease was contracted, but that a gradual process was the cause. Therefore the claim was disallowed.[1]

In some cases the courts have held that disease of an employee contracted in the course of his employment and arising out of it, and due to impure air in a mine, occasioned by the negligence of the employer, and not caused by traumatic injury, was not compensable under the workmen's compensation act, but for such disease the employee may have an action at common law.[2] In an Iowa case the court on the other hand, held as follows:

If an employer fails to provide a reasonably safe place to work, or fails to observe the specific requirements of the statute with respect thereto, and as a result of such negligence the employee is injured, the liability of such employer cannot be avoided by calling such an injury an occupational disease, or by showing that disease of that nature is often the accompaniment or result of such employment.[3]

A worker in Texas set up the claim that due to the inhala-

[1] Rosenthal *v.* National Aniline and Chemical Co., 215 N. Y. Supp. 621, 216 App. Div. 588 (1926).

[2] Jellico Coal Co. *v.* Adkins, 197 Ky. 684, 247 S. W. 972 (1923); Elkhorn Coal Corp. *v.* Kerr, 203 Ky. 804, 263 S. W. 342 (1924); Midland Coal Co. *v.* Rucker's Administrator, 211 Ky. 582, 277 S. W. 838 (1925).

[3] Gay *v.* Hocking Coal Co., 184 Iowa 948, 169 N. W. 360 (1918).

tion of poisonous gases and fumes in the regular course of his employment he developed an occupational disease for which he claimed compensation. The court held that the compensation act was confined in its operation to accidental injuries, and as the plaintiff's disease developed gradually, the accidental element was eliminated, and plaintiff was not entitled to recover under the law.[1]

### LEAD POISONING

Mention has previously been made of the difficulties involved in determining when lead poisoning has actually occurred, and also, whether lead absorption is to be considered as lead poisoning. This lack of standardization of terminology in connection with a disease entity which of necessity presents real problems in diagnosis, must unavoidably cause considerable difficulty for those charged with the administration and interpretation of workmen's compensation laws.

In a Massachusetts case the court held that within the workmen's compensation act, the term "personal injury" included any injury or disease, as lead poisoning, arising in the course of employment, causing incapacity for work and thereby impairing the ability for earning wages. A worker employed as a lead grinder, continuously absorbing lead into his system, receives a personal injury, only, however, when the accumulated effect thereof manifests itself and he becomes sick and unable to work.[2]

Where a common laborer, not in the course of his ordinary occupation, was, by the negligence of his employer, subjected to lead poisoning, from the effects of which he died,

---

[1] Gordon *v.* Travelers Insurance Co., Tex. Civil App., Oct., 1926, 287 S. W. 911.

[2] *In re* Johnson, 217 Mass. 388, 104 N. E. 735 (1914); O'Donnell's Case, 237 Mass. 164, 133 N. E. 621 (1921); Bergeron's Case, 243 Mass. 366, 137 N. E. 739 (1923).

the lead poisoning was held not to be an occupational disease, but it was a " personal injury sustained in the course of employment." [1]

Under the Oregon law providing compensation for a personal injury by accident caused by violent or external means, it was held that the element of suddenness or precipitancy was essential to an accident. An occupational disease such as lead poisoning, caused by the practice of putting tacks in the mouth, was not an accident for which compensation could be granted. [2] In a Michigan case the court similarly denied compensation for lead poisoning, on the ground that occupational diseases were not within the scope of the workmen's compensation law. [3]

### PHOSPHORUS POISONING

A worker employed in a fireworks factory developed phosphorus poisoning. The court held that the Workmen's Compensation Act of Maryland was not restricted in operation to injuries arising from accidents, but included every injury which can be suffered by a worker in the course of and arising out of employment. In holding that phosphorus poisoning was compensable the court stated that the law indicated not the existence of an accident, but rather the idea that the injury was unexpected or unintended. [4]

---

[1] Roth *v.* Industrial Comm. of Ohio, 7 Ohio App. 386, 120 N. E. 172 (1918).

[2] Iwanicki *v.* State Ind. Acc. Comm. of Oregon, 104 Ore. 650, 205 Pac. 990 (1922).

[3] Adams *v.* Acme White Lead and Color Works, 182 Mich. 157, 148 N. W. 485 (1914).

[4] Victory Sparkler and Specialty Co. *v.* Francks, 147 Md. 368, 128 Atl. 635 (1925).

### TUBERCULOSIS

Tuberculosis has long been recognized as one of the major diseases among the industrial workers. Despite the rapid decline during recent years in deaths from pulmonary tuberculosis in the United States, it is still one of the chief causes of death among those of working age.[1] To date, there is but little help that the tuberculous workers can expect through workmen's compensation laws. Many cases have come before compensation boards and the courts with claims arising out of the development of pulmonary tuberculosis among industrial workers. Such claims may in general be divided among those arising out of direct injury accelerating a pre-existing latent or active condition of tuberculosis, and those which are the direct result of the nature of the employment. Instances of both groups follow.

A worker was engaged to varnish a drum. Upon opening it an explosion of gas followed because of gas leakage. The worker suffered severe burns for which he was treated for a number of weeks. Pains in the chest developed shortly after the explosion, and, after nine weeks, death resulted from miliary tuberculosis. Medical testimony was to the effect that gas fumes would furnish an opportunity, if infection existed in a latent form, for such to be kindled into an active condition, and that if infection did not exist in fact, the explosion would bring about destruction of the air cells in the lungs, and make one more susceptible to infection. The court held that the evidence of the deceased's good health before the accident and his state of health after the injury, together with the medical testimony, furnished a sufficient basis to warrant the con-

---

[1] For decline in tuberculosis rate see: *Statistical Bulletin*, Metropolitan Life Ins. Co., vol. x, no. 12, Dec., 1929, pp. 1-4. For figures indicating high death rate in working ages 15-65 years for males and 15-54 years for females see: N. Y. City Dept. of Health, *Ann. Rept.*, 1928, p. 144.

clusion that the miliary tuberculosis was caused by the gas explosion.[1]

A carpenter was employed by a coal company and in the course of his employment sustained an injury to his head and abdomen. For some time thereafter he was ill, and subsequently developed tuberculous peritonitis, from which he died about eight months after the injury. The court held that: " The injuries sustained by the decedent were due to violence to the physical structure of his body, which so lowered his vital resistance that a tuberculous condition quickly developed, hastening and causing his death." On these findings the court granted compensation.[2] Similar cases have involved an injury to the ankle of the worker, resulting in tuberculosis;[3] injury to the back with subsequent pulmonary tuberculosis which had been lighted up by the accident, the court stating: " The evidence authorizes the inference that the accidental injury suffered by the worker aroused the latent germs of the disease to which he was predisposed, materially accelerated the disease, and caused his death earlier than it would otherwise have occurred; "[4] personal injury resulting when a fellow-employee snatched a chair in which the claimant was seated, resulting in a heavy fall to the floor, with subsequent aggravation or acceleration of pre-existing pulmonary tuberculosis;[5] and a number of others in which an injury contributed to, accelerated or ag-

---

[1] Heileman Brewing Co. *v.* Schultz, 161 Wis. 46, 152 N. W. 446 (1915).

[2] Kelly *v.* Watson Coal Co., 115 Atl. 885, Penna. Sup. Ct., Jan., 1922, 272 Pa. 39. See also: Dumbluskey *v.* Phila. and Reading Coal and Iron Co., 270 Pa. 22, 112 Atl. 745 (1921); Miller *v.* Director General of Railroads, 270 Pa. 330, 113 Atl. 373 (1921).

[3] Kelly *v.* Watson Coal Co., 272 Pa. 39, 115 Atl. 885 (1922).

[4] Retmier *v.* Cruse, 67 Ind. App. 192, 119 N. E. 32 (1918).

[5] Healey's Case, Sup. Jud. Ct. Maine, Sept., 1924, 126 Atl. 21.

gravated an already existing condition of pulmonary tuberculosis, and for which the courts allowed compensation.[1]

Important decisions have been rendered in several cases in which pulmonary tuberculosis arose, accidentally or otherwise, out of the nature of the employment. Death from tuberculosis caused by aniline poisoning through " minor faults in equipment or accidental mishandling of materials " has been held to be due to an injury to the lungs in the course of employment and not to an occupational disease; [2] a tuberculous condition of a foundry employee due to breathing fumes and gas when trapped in a furnace by a falling door, was held to be compensable.[3]

A worker was engaged for 16 years as a sand buffer. The work caused minute particles of sand to be thrown into the atmosphere, as a result of which the employee developed pulmonary tuberculosis, which incapacitated him until his death. The court held that the conditions under which the decedent worked were the direct and proximate cause of the tuberculosis and consequent death, and granted compensation to the dependents, because of the personal injury arising out of the employment.[4] In a similar case decided by the same court, it was held that an injury is compensable when tuberculous infection occurs, even though it is not traceable to a definite occurrence at certain times.[5]

[1] McGoey v. Turin Garage and Supply Co., 186 N. Y. S. 697, 195 App. Div. 436 (1921); Gibb v. New Field By-Products Coal Co., 287 Pa. 300, 135 Atl. 207 (1926); Glennon's Case, 236 Mass. 542, 128 N. E. 942 (1920); Republic Iron and Steel Co. v. Markiowicz, 75 Ind. App. 57, 129 N. E. 710 (1921).

[2] Burckard v. Industrial Comm., 22 Ohio L. R. 420 (Comm. Pleas), (1924).

[3] Industrial Comm. v. Rice, 26 Ohio App. 497, 160 N. E. 484 (1927).

[4] Mesite v. International Silver Co., 104 Conn. 724, 134 Atl. 262 (1926).

[5] Kovaliski v. Collins Co., 102 Conn. 6, 128 Atl. 288 (1925); Cishowski v. Clayton Mfg. Co. et al., 105 Conn. 651, 136 Atl. 472 (1927); Dumbrowski v. Jennings and Griffin Co., Sup. Ct. of Errors of Conn., Jan., 1926, 131 Atl. 745.

The Wisconsin court in 1924 ruled that where the conditions under which the occupation of granite cutting must be carried on are such that the consequent filling of the lungs with granite dust necessarily makes one so employed particularly susceptible to pulmonary tuberculosis, it is an occupational disease within the meaning of the workmen's compensation act.[1]    In a Kansas case it was held that where a quarry employee who had worked three years in a dusty department of a cement plant, shortly after striking a large rock with a 16-pound sledge suffered a pulmonary hemorrhage, and died before aid could reach him, death was due to an injury by accident raising out of the employment.[2]

A series of cases have been decided over a period of years in which cases similar to some of those cited above have been denied compensation.    An employee about a coke plant engaged in scraping out offtake pipes through which gas was discharged, was made ill by exposure to the gas.    He subsequently developed tuberculosis, but it was held that he suffered no injury from an accident and therefore was not entitled to compensation.[3]

A worker was employed where a solution of nitric and sulphuric acid was used in the manufacturing process, with resultant hazardous gases and fumes.    Through the negligence of a fellow employee, the plaintiff on a given day was overcome by these poisonous gases and fumes.    From that date he suffered from respiratory ailments, till three years later it was found that pulmonary tuberculosis had set in as a result of the accident.    The court held that compensation could not be granted for the tuberculosis.[4]

[1] Wenrich v. Warning, 182 Wis. 379, 196 N. W. 824 (1924).

[2] Gilliland et al. v. Ash Grove Lime and Portland Cement Co., 104 Kan. 771, 180 Pac. 793 (1919).

[3] Clinchfield Carbocoal Corp. v. Kiser, 139 Va. 387, 124 S. E. 271 (1924).

[4] Campbell v. Industrial Comm. of Ohio, Ohio Ct. of Appeals, 153 N. E. 276 (1926).

A worker in a plant manufacturing shoe polish was employed to stir and bail various chemicals. Certain gases, fumes and powders, all more or less poisonous and irritating, were stirred up in the process of bailing and were inhaled by the worker. This exposure resulted in pulmonary tuberculosis and death. The court held that death was due to an industrial or occupational disease, and not to an accidental injury, and therefore the claim for the death was not compensable.[1]

In a New York case, it was held that tuberculosis arising from work in a damp, unsanitary and unventilated cellar, or inadequately heated workplace was not an accidental injury within the meaning of the compensation law and compensation was denied.[2] Lessened resistance to tuberculosis caused by working in a room containing vapors from acids has been held not to be compensable as an " accident." [3]

### MISCELLANEOUS

A comparatively large number of cases have been decided in which claims for presumable occupational diseases have been set up, with compensation being granted and denied in various states. Some of these cases will be briefly noted. Loss of eye-sight by a moving-picture operator due to ultra-violet rays was held not to be the result of an injury sustained in the course of employment;[4] a case of leather poisoning as a result of handling wet leather was denied compensation and held not to be an injury caused by an acci-

[1] Aetna Life Ins. Co. *v.* Graham, Tex. Appeal Comm. Sec. B, 1926, 284 S. W. 931.

[2] Wager *v.* White Star Candy Co., 217 N. Y. S. 173, 217 N. Y. App. Div. 316 (1926).

[3] Depre *v.* Pacific Coast Forge Co., 145 Wash. 263, 259 Pac. 720 (1927).

[4] Industrial Comm. of Ohio *v.* Russell, 111 Ohio 692, 146 N. E. 305 (1924).

dent;[1] muscular paralysis, due to exposure to cold while handling cold articles in a refrigerator was held not to be an accidental injury;[2] cases of heat-stroke and sun-stroke have been compensated on the ground that an accidental injury had occurred;[3] infection not accidental in nature but resulting from continuous pressure of hand tools has been held not to warrant a finding of injury by accident.[4] Scalds and burns resulting from the regular use of soda ash were held to be compensable injuries;[5] empyema developing slowly and gradually from continued contact with smoke and fumes was held to be an occupational disease and not compensable;[6] where a carpenter developed " housemaid's knee " from being on his knees scraping and polishing a floor a few days before he was disabled, it was held that the injury, even if an occupational disease, was an accident arising out of and in the course of employment within the meaning of the Indiana Workmen's Compensation Act, as the word " accident " was held to mean an unlooked for mishap, or untoward event not expected or designed.[7]

[1] Dillingham's Case, Sup. Jud. Ct. Maine, Aug., 1928, 142 Atl. 865.

[2] Chop *v.* Swift and Co., 118 Kan. 35, 233 Pac. 800 (1925).

[3] Matis *v.* Schaeffer, 270 Pa. 141, 113 Atl. 64 (1921); Walsh *v.* River Spinning Co., 41 R. I. 490, 103 Atl. 1025 (1918); State *ex rel.* Rau *v.* District Ct. of Ramsey Co., 138 Minn. 250, 164 N. W. 916 (1917); Young *v.* Western Furniture and Mfg. Co., 101 Neb. 696, 164 N. W. 712 (1917).

[4] Perkins *v.* Jackson Cushion Spring Co., 206 Mich. 98, 172 N. W. 374 (1919).

[5] Ward *v.* Beatrice Creamery Co., 104 Okla. 91, 230 Pac. 872 (1924).

[6] Manchline *v.* State Ins. Fund, 279 Pa. 524, 124 Atl. 168 (1924).

[7] Standard Cabinet Co. *v.* Landgrove, 76 Ind. App. 593, 132 N. E. 661 (1921).

### CONCLUSION

The preceding analysis of judicial decisions affecting cases of occupational diseases, is perhaps indicative of the complexity of the present system of handling such cases, and of some of the difficulties involved. As noted, but few of the states provide for compensation, and the tendency is for such as have schedules of diseases to add to such schedules from time to time. Thus, during 1929 and in 1930 additions were made in New York,[1] and in 1929 in Ohio.[2] A careful analysis of industrial processes in this country has produced a list of 700 hazardous occupations in which there are abnormalities of working conditions.[3] As the diagnosis of many of the resultant diseases is frequently a matter of considerable medical and technical difficulty, it becomes apparent that further additions to occupational disease schedules may even tend to complicate the present unsatisfactory situation.

The courts are beset with many problems in adjudicating occupational disease cases that come to them on appeal. The laws covering compensation for such diseases may not be sufficiently explicit; medical and other expert testimony may be conflicting; differential medical diagnosis may be involved; diagnosis of newly discovered occupational diseases may present various difficulties—these and other matters tend to make the task of the courts quite complex. It oftimes becomes virtually impossible for the courts in different jurisdictions to decide such cases with justice to both the employer and the employee.

Some other method of compensating workers for occupational diseases might seemingly be more desirable than de-

[1] *Consol. Laws*, amended 1929, ch. 298; 1930, ch. 60.

[2] *Gen. Code*, secs. 1465-1468a, as amended 1929.

[3] Dublin, Louis I., " Occupation Hazards and Diagnostic Signs," *U. S. Bur. Lab. Stat. Bull. no. 306, 1922.*

pendence upon so-called blanket and schedule plans.    The appeals to and the decisions of the courts, with some of the attendant problems already enumerated, as well as cost to the injured worker in legal expenses, delay, etc., further complicate the situation.    Such other method or procedure may possibly be found in some system of health insurance, which will be dealt with in the following pages.

# CHAPTER VI

## HEALTH INSURANCE FOR OCCUPATIONAL DISEASES

### INTRODUCTION

It has been recognized for some time that when the people of an entire state are subjected to certain risks which are measurable, it is good business to organize some kind of insurance plan through the instrumentality of the state, to measure the risk, and to pay the losses which happen out of a central fund. In this way there is a distribution of the risk over a wide area and involving many people, making it possible to cover losses which may occur without undue hardship to particular individuals. This is the idea behind all successful insurance plans. These plans are already in effect in a number of states, covering such matters as insurance of bank deposits, hail insurance and workmen's compensation insurance.[1]

The importance of adequate provision in case of illness or invalidity was recognized by workers long before the advent of health insurance. As early as the Middle Ages the insufficiency of individual action was realized, and a more satisfactory arrangement, that of insurance, was initiated by the guilds. Insurance was purely voluntary and the workers had to bear the full cost.[2] These guilds, to which belonged masters and journeymen of the same craft, often provided for the assistance of their members in case of sickness and

---

[1] Lapp, John A., *U. S. Bur. Lab. Stat. Mo. Lab. Rev.*, vol. ix, no. 9, Sept., 1919, pp. 938-944.

[2] Commons and Andrews, *op. cit.*, p. 416.

injury. Such a plan of sickness insurance continued in some form until the abolition of the guilds by Great Britain in 1835.[1]

As the dangers of sickness and the economic distress consequent upon it have existed in many forms of industrial organization, the beginnings of sickness insurance may be traced further back than those of insurance against accidents, or any other form of social insurance. These beginnings are to be found in the organization of mutual aid similar to the plan of the guilds, though not necessarily limited to any special social group. The early plans were of a purely voluntary form of organization, and were supported exclusively by the contributions of the members.[2]

### DEVELOPMENT OF HEALTH INSURANCE IN GERMANY

The first country to make sickness insurance compulsory was Germany, and it was Bismarck who contributed to the history of social insurance the first application of state compulsion on a large national scale.[3] This creation of Bismarck's was keenly opposed at first by the German working class, and was presumably inspired by a political motive. It aimed at rendering innocuous the numerous relief funds connected with trade unions, which might have been dangerous weapons in a class struggle. By making secure the means of subsistence of the working class, his plan raised a barrier against Socialist propaganda. But whatever may have been the intentions of its originator, the German scheme has exercised a considerable influence on the legislation of a number of other countries.[4]

[1] International Labour Office, *Studies and Report*, Series M., no. 7, "Voluntary Sickness Insurance," 1927, pp. xix, xx.

[2] U. S. Commissioner of Labor, *24th Annual Report*, 1909, "Workmen's Insurance and Compensation Systems in Europe," vol. i, p. 7.

[3] Rubinow, I. M., *Social Insurance* (New York, 1913), p. 13.

[4] International Labour Office, Series M., no. 8, pp. 10-11.

Germany did more in erecting this system than accomplish a great reform for the welfare of its people. It set an example that has powerfully influenced many industrial nations. Scarcely a country in Europe but immediately applied itself to the task of studying the system, of examining its own needs in this field, and of elaborating propositions with a similar end in view.[1]

The principle of mutual insurance in Germany was, as already indicated, by no means a new one. In South Germany there was a system in the parishes of sick relief of servants and dependent laborers, which was based upon the insurance principle. In addition, there were also in the north many benefit societies of different kinds. The benefit societies which were prevalent in the years just before Bismarck's measures came into effect, and upon which he based his own plans, were to be found among the miners in the Harz Mountains and elsewhere. These " Knappschaftskassen," as they were called, had existed since early in the 16th century. At first they were entirely in the hands of the workpeople, who bore the full cost. The men of each mine were, as a rule, separately organized, with rates of contribution and relief varying considerably. As time went on the system was more and more unified. The employers began to cooperate with the workers, with contributions from both groups, and subsequently, membership in these sickness funds became compulsory.[2]

The draft of the first sickness insurance bill, which was announced as early as 1881 became law, with some amendments, on June 15, 1883. In the course of the following years the circle of persons liable to insurance was repeatedly

[1] Willoughby, Wm. F., " The Problem of Social Insurance," *Amer. Labor Legis. Rev.*, vol. iii, 1913, p. 154.

[2] Ashley, Annie, *The Social Policy of Bismarck* (London, 1912), pp. 46-47.

enlarged, and the extent of the benefits steadily increased. In 1911 the Federal Insurance Code (Reichsversicherungs-ordnung) brought together the Sickness Insurance Act and the more recent acts for protection against invalidity and accidents in a single text.    The desired assimilation and uni-fication in the granting of benefits was not achieved.    A special Salaried Employees' Insurance Act, protecting them against invalidity, old age and death, was passed in the same year, and this act remained distinct from the Federal Insur-ance Code.    The Federal Insurance Code and the Salaried Employees' Insurance Act were re-enacted in new form on December 15, 1924 and May 28, 1925 respectively.

Shortly before this, on June 26, 1923, all German miners' benefit societies had been combined into one body, the Fed-eral Miners' Benefit Society (Reichsknappschaftsverein), whereby all the trades engaged in mining operations achieved a single system of sickness, invalidity and old age insurance. Insurance against accidents was left as a part of the general organization. Fresh modifications and supplementary measures having been introduced, the Federal Insurance Code was re-enacted on January 9, 1926, though the entire revision of the system was left uncompleted.[1]

A recent report indicates that approximately 20,288,000 people in Germany are insured against illness in sick-clubs under the insurance laws.    The dependents of insured per-sons also enjoy some of the benefits, and if they are included, it may be said that about two-thirds of the German popula-tion belong in some way or another to a sick-club.    There are 7,679 institutions of this kind under various descriptions in Germany, the largest being in Berlin with over 484,000 members.    There are moreover, a number of clubs for

---

[1] Goldman, Franz, and Grotjahn, Alfred, *Benefits of the German Sick-ness Insurance System*, Inter. Labour Office (Geneva, 1928), pp. 4-5.

special trades.[1]  The extent of the medical service is indicated, for example, in the report for 1925, during which year the Dresden General Local Sickness Fund had 123,000 visits; Leipzig 300,000, and Nuremberg 284,000 visits.[2]

### DEVELOPMENT OF HEALTH INSURANCE IN ENGLAND

In England, until recent date, "social insurance" was better known as "workingmen's insurance" or "labour insurance".  But the term "social" has gradually come into use in recognition of the fact that even outside of the wage-earning class there are certain elements of the population who need protection and who are included under the folds of this legislation.[3]

The mutual aid funds which started in Great Britain embodied the workers' first attempts to provide against sickness risks, but the field of these organizations was limited, and the resources too small to enable them to undertake insurance against prolonged illness or disablement.[4]  At the outset mutual aid societies were based on trades, like the medieval guilds.  While, however, a society's connection with the trade association, from which it sprang, secured for it the support of trade unionists, the range of its membership was at the same time restricted, and its financial resources were limited to the small contributions which the organized workers were able to pay.  Its basis of organization was not large enough to enable it to undertake insurance against the more serious risks of prolonged illness or disablement.  Nevertheless, it supplied the foundation of a system which was later to be developed with the aid of public authorities.[5]

[1] *The Lancet*, no. 5487, London, Oct. 27, 1928, p. 890.

[2] Goldman and Grotjahn, *op. cit.*, pp. 153-154.

[3] Cohen, Joseph L., *Social Insurance Unified* (London, 1924), p. 25.

[4] *U. S. Bur. Lab. Stat. Mo. Lab. Rev.*, vol. xxvi, no. 6, June, 1928, p. 1171.

[5] International Labour Office, Series M, no. 6, *Compulsory Sickness Insurance*, 1927, p. 8.

Great Britain built its insurance system around the voluntary friendly societies, utilizing their organization and permitting them to establish separate sections for national insurance.[1]  Moreover, the British plan was considerably influenced by the popularity of the large friendly societies, which became the main vehicles of the plan of sickness insurance.[2]  It is probable that the National Insurance Act was indirectly the outcome of the Report of the Royal Commission on the Poor Laws (1907-1909).  Both the Majority and Minority Reports called attention to the association of poverty and sickness, but neither recommended a health insurance plan as a remedy.[3]  The reports favored the abolition of the Boards of Guardians which dispensed poor relief.  The Maclean Report of 1917, with practically entire unanimity, endorsed the Royal Commission's recommendation that the Boards of Guardians should be abolished. There has been a definite stand taken by proponents of the health insurance plan that the Boards of Guardians represented an archaic system embodied in the Poor Law, and that, for the future development of health insurance and the national welfare, it would be best to follow the recommendations of the Royal Commission of 1907-1909.[4]

Lloyd George, in a speech in the House of Commons on May 4, 1911, in introducing the bill on social insurance, presented many arguments in its favor.  He stated among other things: " In this country 30 per cent of the pauperism is attributable to sickness.  A considerable percentage would probably have to be added to that for unemployment." [5] He further pointed out that the bill was intended to effect

---

[1] Commons and Andrews, *op. cit.*, p. 421.

[2] Rubinow, I. M., *Social Insurance*, p. 25.

[3] Brend, Wm. A., *Health and the State* (London, 1917), p. 211.

[4] Gordon, Alban, *Social Insurance* (Westminister, 1924), pp. 66-67.

[5] Lloyd George, David, *The People's Insurance* (London, 1911), p. 1.

as wide an insurance as possible of the working population against sickness and breakdown, and to make the bill a preventive measure operating to reduce the amount of sickness. After indicating that the Majority and Minority Reports of the Poor Law Commission called special attention to the utter inadequacy of methods for preventing and curing sickness among the industrial classes, he noted that the bill contained several provisions designed to amend this unsatisfactory state of things.[1]

National health and invalidity insurance was placed in operation on a compulsory basis in 1912, when the Act of December 16, 1911, came into force. All workers between the ages of 16 and 70 were placed within the scope of the act. It was estimated at the time that there would be about 9,-200,000 men and 3,900,000 women who would be subject to the health insurance system. The actual experience later showed that these estimates were approximately correct, as the number of men insured in 1914 was about 9,680,000 and the number of women about 4,077,000, or a total of about 13,757,000.[2]

Prior to the National Health Insurance plan, contract practice of physicians reached a large number of wage earners and their families.[3] For many years, physicians in industrial and agricultural areas recognized that there were considerable sections of the public who would never receive or could not pay for medical attendance unless some arrangement was made on a contract basis. Many physicians had their own " sickness clubs ", generally collecting from their

---

[1] *Ibid.*, p. 35.

[2] *U. S. Bur. Lab. Stat. Bull. no. 312*, "National Health Insurance in Great Britain 1911 to 1921," April, 1923, p. 1.

[3] Lloyd George, David, *The People's Insurance*, states that between 6,000,000 and 7,000,000 people had made some provisions against sickness, not all of it adequate, and a good deal of it defective, pp. 3-4.

members a small weekly contribution for which the whole family was attended. The most widespread contract practice arrangement was through the friendly societies and trade-unions. The members of these organizations paid weekly contributions. Each local branch or " lodge " would appoint its own doctor and pay him a fixed sum, for which he undertook to give ordinary medical attendance and medicines to each member.[1]

From the very start it was recognized that the plan would have to be modified from time to time to adjust it to conditions not foreseen when the plan was prepared, or to improve it as experience showed the necessity for such change. The principal amending laws have been those of 1913, 1918 and 1920. The first two of these provided for simplifying the administration and correcting certain abuses; the 1920 amending act increased the contributions and benefits, and provided for the eventual transfer of the tuberculosis sanatorium benefit to a new service in order to supply a more comprehensive course of treatment.[2]

General comments upon the results of the British system of health insurance might be condensed into the summary that the Health Insurance Act gave some of the insured persons a service they had been providing for themselves privately; it gave a large number a service they had never been able to get before except partially and occasionally through charity; it gave a much larger number a service similar to, though better than, what they had been getting on a contract basis.[3]

The Royal Commission on National Health Insurance, in the report submitted in 1926, stated as follows:

[1] Cox, Alfred, *Seven Years of National Health Insurance in England* (Chicago, 1921), pp. 4-5.

[2] *U. S. Bur. Lab. Stat. Bull. no. 312*, p. 3.

[3] Cox, Alfred, *op. cit.*, p. 16.

We can say that we are satisfied that the scheme of National Health Insurance has fully justified itself and has, on the whole, been successful in operation. The workers of this country have obtained under it substantial advantages, in particular by securing the title to free medical attention and medicine whenever and as soon as these are required, and by the proportionate diminution, to the extent of the cash benefit granted, of their anxiety as to the loss of wages during illness.

We are convinced that National Health Insurance has now become a permanent feature of the social system of this country, and should be continued on its present compulsory and contributory basis.[1]

### COMPARISON OF GERMAN AND BRITISH SYSTEMS

The following indicates the fundamental features of the Federal Insurance Code of Germany, insofar as provision is made for sickness insurance, and the main features of the British health insurance plan.

*Economic Activities Covered by Compulsory Sickness Insurance* [2]

*Germany*—industry, commerce, agriculture, navigation, railways (manual workers only [3]), domestic service, home workers.

*Great Britain*—industry, commerce, agriculture, navigation, railways (manual workers only [4]), domestic service, home workers.

---

[1] *Report of Royal Commission on National Health Insurance* (London, 1926), p. 16.

[2] International Labour Office, *Compulsory Sickness Insurance*, Series M, no. 6, pp. 154-155.

[3] Salaried employees of railways, civil servants and certain employees of local authorities, entitled to equivalent treatment, are excluded.

[4] Salaried employees and civil servants, entitled to equivalent treatment, are excluded.

### Workers Excluded from Scope of Compulsory Sickness Insurance [1]

*Germany*—invalids at own request; non-manual workers earning more than 3,600 Marks a year; members of employer's family working without agreement or remuneration; employment for less than one week of persons not ordinarily engaged in paid work.

*Great Britain*—under 16 years and for cash benefits over 65 years; non-manual workers earning more than £250 a year; wife or husband of employer; child employed without payment; persons casually employed otherwise than for purposes of employers' trade; certain seasonal agricultural workers.

### Conditions for Award of Statutory Sickness Benefit [2]

*Germany*—no qualifying period of insurance; day of incapacity as from which benefit begins to be paid is fourth day.

*Great Britain*—qualifying period of insurance is 26 weeks; day of incapacity as from which benefit begins to be paid is fourth day.

### Amount and Duration of Sickness Benefit [3]

*Germany*—rate of benefit payable to insured person incapable of work is 50 per cent of basic wage; rate payable to dependents of hospital inmate is 50 per cent of statutory benefit; duration of statutory benefit 26 weeks.

*Great Britain*—rate benefit 15 s. for men, 12 s. for women; to dependents of hospital inmate, whole or part of statutory benefit at discretion of fund. Duration of statutory benefit 26 weeks.

### Share of Contribution [4]

*Germany*—insured person ⅔; employer ⅓; no share toward cost of administration.

[1] *Ibid.*, pp. 156-157.
[2] *Ibid.*, p. 397.
[3] *Ibid.*, p. 398.
[4] *Ibid.*, pp. 430-431.

*Great Britain*—insured men ½; employer ½; insured women 8/17; employer 9/17; cost of central administration covered by government.

### Medical Benefits for Dependents [1]

*Germany*—compulsory for wife and children of miners; medical and hospital treatment; 50-70 per cent of cost of drugs; maximum duration 26 weeks.
*Great Britain*—no compulsory benefits.

#### SICKNESS INSURANCE IN OTHER COUNTRIES

The European states created by the peace treaties have passed laws amending the insurance schemes they inherited. Among them have been Czechoslovakia in 1919; Poland in 1920; Austria in 1921; in 1922 Yugoslavia made sickness insurance compulsory for all wage earners; Portugal in 1919; Greece in 1922 and Bulgaria in 1924 passed laws to cover all classes of workers.[2]

Soviet Russia abandoned the system of public assistance established in 1918 for compulsory insurance, the plan of which was incorporated in the 1922 Labor Code. Lithuania in 1925 passed a sickness insurance law, but it has not as yet been enforced. Countries outside of Europe which have adopted compulsory systems are Japan (1922) and Chile (1924), while the governments of Australia and South Africa have appointed commissions to study schemes of compulsory social insurance. Brazil is preparing a labor code which will include compulsory insurance against sickness.[3]

The social insurance law which was enacted by the French Parliament March 14, 1928, and promulgated April 5, 1928,

---

[1] *Ibid.*, p. 401.

[2] *U. S. Bur. Lab. Stat. Mo. Lab. Rev.*, vol. 26, no. 6, June 1928, p. 78.

[3] International Labour Office, *Compulsory Sickness Insurance*, Series M, no. 6, 1927, p. 11.

but which was never put into effect owing to the objections of various groups, has been amended in several important particulars. The new act which became effective July 1, 1930, provides for two distinct systems—one for commercial, industrial and domestic workers, and the other for agricultural workers. The insurance system does not include civil servants, miners, railway workers, seamen, etc., who are already protected by special legislation. However, a decree will be issued before July 1, 1931, fixing the rules for the coordination of these various systems with the general social insurance scheme. Under the new plan, all wage-earners are compulsorily insured if their wages do not exceed 15,000 francs, or 18,000 francs in cities of more than 200,000 inhabitants. The insurance system is financed by equal contributions by the employer and the worker, supplemented by certain contributions by the state. Insured persons receive both general and special medical care, medicines and appliances, treatment in hospitals or sanatoria, necessary surgical operations and preventive treatment. Such cases of illness as are due to occupation or industry are cared for under the workmen's compensation laws. An insured person has free choice of a physician. The plan is in effect but a few months and will require some years of adjustment to French conditions to make it more effective in its application than it can possibly be after so short a trial period.[1]

## HEALTH INSURANCE IN THE UNITED STATES

### *Introduction*

A transformation of industry similar to that in Great Britain and Germany has been taking place in the United States for some time. The questions that naturally arise

[1] *U. S. Bur. Lab. Stat. Mo. Lab. Rev.*, vol. 31, no. 3, Sept., 1930, pp. 76-78.

are, whether or not the development of health insurance legislation in the United States on the model of the British and German laws would be desirable; and further, would it be advisable to substitute a system of compulsory social insurance for individual and voluntary provision against sickness.[1]

It should be recognized that social insurance seeks not merely that provision shall be made for all cases where the worker is unable through illness and by his current labor to support himself and those dependent upon him, but that the burden of this provision shall be placed where it equitably belongs. It seeks further to prevent this burden from falling upon individuals and to place it upon groups, where it can most readily be borne. Further, it aims to improve the mechanism by which payments due because of occupational diseases or other illness, are determined by taking this matter from the domain of litigation and the courts, and providing a more simple, economical, expeditious and certain machinery and procedure.[2]

Under the present methods of compensating for occupational diseases, the cause of the injury becomes the most important and difficult question when reporting such injury.[3] The justice of including all occupational diseases under workmen's compensation laws is rapidly being recognized, though the costs to the employers have frequently been found to be a cause of opposition. The difficulty, in the vast number of cases, is to prove, under the compensation laws, that the disability is due to the nature of the employment and not to something else.

Some have regarded an infection as an acute form of oc-

---

[1] Woodbury, R. M., *Social Insurance, An Economic Analysis* (New York, 1917), p. 21.

[2] Willoughby, Wm. F., "The Problem of Social Insurance," *Amer. Labor Legis. Rev.*, vol. iii, 1913, p. 161.

[3] Andrews, John B., *Amer. Labor Legis. Rev.*, vol. viii, 1918, p. 314.

cupational disease, while others consider it as an accident. That is to say, certain experts look upon infection in itself as a sudden occurrence due to a violent cause (accident); others regard it as the complication of an accident. Finally, a third group treats it as an ordinary disease. In countries not having sickness insurance laws, cases of anthrax, glanders, etc., have been treated as accidents, or rejected as forms of injury involving compliance with or departure from the general rule. The disease brought on by compressed air and lying on the borderline between occupational diseases and accidents, is considered by some as a typical form of occupational disease, while others look upon it as an accident.[1]

The frequently slow progress of occupational diseases makes the placing of the responsibility upon the individual employer (usually the last employer) somewhat a problem. The difficulty of determining where and when an occupational disease was originally acquired is a real one, though the importance of it may be exaggerated. It is still more difficult to determine whether a disease is occupational in origin or not, i. e., whether the particular case was due to the occupation, and whether it was incurred in the place of employment at all. This problem arises because of the lack of a precise and generally accepted definition of the term "occupational disease."[2] The previous condition of the diseased worker constitutes a further problem which arises as soon as an attempt is made to fix the allowance to be granted either in case of accident or in the case of disease, more particularly where certain forms of disease, such as tuberculosis, cancer, etc., are concerned.

Difficult as it may be to arrive at an exact estimation of the influence of the past medical history, and of the condi-

---

[1] International Labour Office, Series M, no. 3, 1925, pp. 20-21.

[2] Rubinow, I. M., *Standards of Health Insurance* (New York, 1916), pp. 56-57.

tions of social life on the subsequent developments of injuries caused by accidents, it is yet more difficult to do so in connection with occupational diseases. Diseases such as pulmonary tuberculosis, neuritis, etc., in a latent condition, which do not prevent the worker from earning his living, may evolve more rapidly and even cause the death of the worker if the noxious effect of a toxic substance, such as lead, or of local conditions of work, as dust, humidity, high temperature, etc., intervene as additional factors. Yet undoubtedly, in many such cases, compensation for disability or death is not granted.[1] The effective and perhaps the just way to handle such matters, is to cover all sickness regardless of its cause, whether due to working conditions or to living conditions and personal habits, and distribute the cost as is done in other countries.

ATTEMPTS AT HEALTH INSURANCE IN THE UNITED STATES

Health insurance for workingmen commenced in this country as a business, in the form of providing hospital service, or its equivalent, cash benefits, chiefly for railway employees. After a setback in Wisconsin, where it began about 1892, it again came into existence in Michigan. From railways it spread to mines which had no benefit funds, or inadequate ones, and to manufacturing plants.[2] A number of methods for providing insurance against illness, such as establishments funds, mutual benefit funds, commercial health insurance, fraternal insurance, and trade-union benefits have been tried in this country.[3] No state, municipal or other governmental division has in any way provided for or aided a health insurance scheme. As far as wage-earners are able, they are rapidly adopting health insurance of their own ac-

[1] International Labour Office, Series M, no. 3, pp. 21-22.
[2] Dawson, Miles M., *Amer. Labor Legis. Rev.*, vol. vii, 1917, p. 113.
[3] *U. S. Bur. Lab. Stat. Bull. no. 212*, p. 327.

cord. A constantly growing number of employers are in-augurating establishment funds which are largely supported by the employees.[1] Commercial companies are finding that their health insurance policies are on the increase.[2]

Existing agencies for health insurance in the United States are usually classified as follows: (1) commercial; (2) fra-ternal; (3) establishment; and (4) trade-union. A fifth type seems to be discernible in the industrial benefit societies which are patterned closely after European models, but which have never attained any considerable growth in this country. These societies, however, are very similar to establishment and trade-union benefit organizations in their methods and differ from them principally in the scope of their member-ship, being confined neither to any one trade or industry, nor to any one establishment.[3]

Trade-unions, through voluntary action, have made at-tempts to provide forms of health insurance. While their plans may be regarded as purely experimental and have proven, in most instances, inadequate, yet they have been productive of some good. But the burden of taking care of workers who are ill and providing adequate hospital and

[1] Sydenstricker, E. and Warren, B. S., "Health Insurance in Its Rela-tion to the Public Health," *U. S. Public Health Bull. no. 76*, March, 1916, pp. 51-52. Chase,Stuart, *Laid Off at Forty*, Harpers, Aug., 1929, notes that about 6,000,000 workers are included in group insurance plans; with this growth the position of the older employee becomes harder be-cause the older the average age of the factory or office staff, the higher are the premiums under the group insurance schedules, p. 344.

[2] Wright, Wade, "Insurance for Health," *Survey*, Jan. 1, 1930, notes that group health and accident insurance, not including workmen's com-pensation insurance, covers about 2,000,000 lives. About half of this insurance is carried by one large mutual company, the balance being dis-tributed among a number of other companies; this insurance about doubled in two years, and at the present rate of growth it promises to be one of the outstanding fields of insurance within a very short time, p. 424.

[3] *U. S. Bur. Lab. Stat. Bull. no. 212*, p. 436.

medical service for them has been altogether too great to be borne by these voluntary organizations.[1]

## FAILURE OF VOLUNTARY INSURANCE

The attempts of the workers to provide health insurance for themselves is clearly indicative of their realization of the need for such a type of protection and surety. Only within recent years have they begun to receive the cooperation of employers, and, in many instances, even this cooperation has been of doubtful value.[2] The systems and extent of sickness or health insurance now in effect are wholly inadequate and reach but a small portion of the total working population. Probably the chief reason for the lack of a wider coverage lies in the matter of expense involved, with inability of large bodies of workers to meet costs which are prohibitive to them.[3]

The very fact that only a small minority of insurable wage-earners are at present insured, even after various agencies have been available for a number of years, is evidence that voluntary insurance in this country, just as in European countries, will not meet the needs of the workers.

A definite limitation to the expansion of existing agencies for health insurance lies in their self-imposed restrictions. Medical examinations frequently prohibit for membership those who are in greatest need of the benefits of health insurance and the medical services at low cost. The great mass of the poorly paid workers are in large measure automat-

[1] For list of unions providing sickness and disability benefits, see: U. S. Bur. Lab. Stat. Bull. no. 465, "Beneficial Activities of American Trade-Unions," Sept., 1928, pp. 14-21.

[2] U. S. Bur. Lab. Stat. Bull. no. 212, pp. 469-470.

[3] Pott, Rufus M., "To What Extent is Social or Welfare Insurance Feasible in the United States," U. S. Bur. Lab. Stat. Mo. Lab. Rev., vol. iii, no. 5, Nov., 1916, p. 74.

ically shut out.[1]   The cost of the reliable commercial and fraternal insurance is too great for them to bear.   Workers in occupations in which there is a high disease hazard, especially where the incidence of occupational diseases is marked, are usually specifically barred from commercially managed funds, or effectually barred by the high premium rates they must pay.[2]   Yet these are the workers who need health insurance most.   The policy of leaving to the low-paid wage-earners the responsibility of providing for sickness and safeguarding their health and that of their families, puts upon them a burden which they cannot carry, even in the face of almost certain contingencies of sickness.[3]

Under all voluntary systems the proportion of the insured in a definite labor group is in inverse ratio to its economic status.   Ability and willingness to meet the cost of insurance presupposes the existence of some surplus in the budget and a sufficient cultural status for the appreciation of the advantages of the insurance principle.   The subsidized voluntary system recognizes that for a large proportion of the wage-earners the incomes are such as to make the cost of insurance too heavy a burden, and endeavors to relieve this burden by some type of subsidy.   However, only through a compulsory system does it become possible to shift part of the cost of health insurance upon the employer and upon industry at large.[4]

As a general rule, only those workers in this country who are able to pay all of the cost of carrying insurance can

[1] *U. S. Census of Manufactures for 1925* indicates, for instance, that in 1925 the average annual earnings in all industries in New York State amounted to $1439; (p. 1349) in New Jersey $1352; (p. 1345) in Massachusetts $1211 (p. 1331).

[2] *U. S. Bur. Lab. Stat. Bull. no. 212*, pp. 470-471.

[3] *U. S. Public Health Bull. no. 76*, pp. 45-46.

[4] Rubinow, I. M., *Standards of Health Insurance*, pp. 23-25.

obtain its advantages, and the cost of operation of health insurance, especially in commercial companies, necessitates a premium which contains a substantial profit to the company.[1] As constituted and practiced in the United States, health insurance practically leaves the problem of the low-paid wage-earners virtually untouched. The commercial companies have served a useful purpose in providing such insurance for the better paid workers, but they have not reached the lower paid groups.[2]

AGITATION FOR HEALTH INSURANCE IN THE UNITED STATES

It seems quite natural that subsequent to the extension of health insurance in European countries, and especially in Germany and England, that the subject should begin to receive serious consideration in this country. The first attempt to formulate a plan of compulsory health insurance adapted to conditions in the United States was that of the American Association for Labor Legislation in 1914.[3] The

[1] Woodbury, R. M., *Social Insurance*: estimates that two-fifths of the premiums collected by industrial insurance companies, with weekly premiums, is paid for expenses of securing and managing the business. Also, that a very large number of lapses take place in spite of the weekly system of collections, p. 31. Lloyd George, in his speech on the Health Insurance Bill, also pointed out the large number of lapses of industrial insurance policies. (*The People's Insurance*, pp. 3, 5, 6.)

[2] The report of the Prudential Insurance Co. for 1928 indicates that the immense sum of $283,249,678 in industrial premiums was paid in 1926 to this company alone. In the same year 3,799,797 new policies were issued in the industrial department, amounting on face value to $1,083,478,588. (*The Insurance Almanac*, 1929, New York, pp. 331-332.) The John Hancock Insurance Co. started to issue group accident and health insurance policies in the same year. (*The Insurance Almanac*, 1929, p. 290.) In 1926 the Metropolitan Life Ins. Co. commenced writing monthly premium industrial policies at rates about 15 per cent less than the weekly rates. (*The Insurance Almanac*, 1929, p. 299.) In 1929 the Metropolitan wrote $1,416,638,094 of industrial insurance. (*New York Times*, Jan. 31, 1930.)

[3] *Amer. Labor Legis. Rev.*, vol. iv, Dec., 1914, pp. 595-596.

Association was instrumental in the formation of the Social Insurance Committee, which, with the cooperation of a committee of the American Medical Association, published the first tentative draft of a health insurance bill.[1]

The " Standard Bill " of the American Association for Labor Legislation, as published June 1916, covered all manual employees and all other employees receiving less than $1,200 per annum. It provided for cash benefits of 66 ⅔ per cent of the wages for a period not to exceed 26 weeks in any consecutive twelve months.   Medical benefits included medical, surgical and nursing assistance and treatment; medicines and therapeutic appliances costing not more than $50 in any one year; hospital care, medical and surgical treatment and medicines to dependents.

Employers were asked to contribute ⅖, employees ⅖, and the state ⅕ of the expenses.   Arrangements were provided for the inclusion of non-profit-making insurance carriers, and the organization of local funds.   The medical service was provided for through: (1) a panel of physicians to which all legally qualified physicians might belong; (2) salaried physicians in the employ of the carriers; (3) district medical officers; and (4) a combination of the above methods.   Several changes were made in the bill in 1915, 1916 and 1918.[2]

The bills which have been introduced in the various state legislatures followed in the main the standards for health insurance formulated by the American Association for Labor Legislation.   They usually provided for a cash sickness

[1] For favorable attitude of American Medical Assn., see: *U. S. Bur. Lab. Stat. Mo. Lab. Rev.*, vol. ix, no. 6, Dec., 1919, pp. 345-346.   The report of the Committee of the A. M. A. indicated that the responsibility for sickness was threefold: communal, industrial and individual; and that the individual has heretofore borne nearly all of the burden, whereas industry and the community should be doing their part.

[2] *Illinois Health Insurance Commission Report*, May, 1919, pp. 634-635.

benefit during twenty-six weeks, medical care, maternity benefits and a funeral benefit. The cost is divided equally between worker and employer, while the state bears the cost of central supervision. The insurance is to be carried by mutual, democratically managed associations of workers and employers, with supervision by the state. In addition to the relief measures the bills have usually included certain measures aiming at the prevention of illness.[1]

State legislatures have thus far seemed inclined to caution in dealing with the question of establishing a compulsory health insurance system. In 1915 the " Standard Bill " drawn up by the American Association for Labor Legislation and a committee of the American Medical Association, was introduced in the legislatures of New York, Massachusetts and New Jersey.[2] In 1917, twelve state legislatures considered health insurance bills.[3] The legislatures of several states appointed commissions to investigate the problems involved in a system of health insurance. Among these states have been the following: California,[4] Connecticut,[5] Illinois,[6] Massachusetts,[7] New Jersey,[8] Ohio,[9] Pennsylvania[10] and Wisconsin.[11]

[1] Commons and Andrews, *op. cit.*, pp. 427-428.

[2] *Illinois Health Insurance Commission Report*, pp. 629-630

[3] Beckner, E. R., *A History of Illinois Labor Legislation* (Chicago, 1929), p. 483.

[4] *Laws of 1917*, ch. 312.

[5] *Laws of 1917*, ch. 163.

[6] *Laws of 1917*, ch. 488.

[7] *Laws of 1917*, ch. 130.

[8] *Laws of 1917*, ch. 277.

[9] *Laws of 1917*, p. 520.

[10] *Laws of 1917*, ch. 414.

[11] *Laws of 1917*, J. R. 24, ch. 604.

### CALIFORNIA

The California commission authorized to study the general problem of social insurance, concentrated on health insurance and reported to the legislature in January 1917. Prior to the 1917 commission, Governor Hiram W. Johnson had appointed a social insurance commission in 1915,[1] following a presentation to the legislature of the widespread problems of dependency and destitution.

The latter commission spent considerable time and effort in fact finding, and in seeking and obtaining the services and advice of those expert in matters pertaining to the problem of health insurance. The commission refrained from submitting a draft of a bill to cover the subject, but drew up a proposed constitutional amendment, which reads as follows:

It is hereby declared to be the policy of the State of California to make special provision for the health and welfare of those classes of persons, and their dependents, whose incomes, in the determination of the legislature, are not sufficient to meet the hazards of sickness. The legislature may establish a health insurance system, applicable to any or all such persons, and for the financial support of such system may provide for contributions, either voluntary or compulsory, from such persons, from employers, and from the state by appropriations.

The legislature may confer upon any commission or court, now or hereafter created, such power and authority as the legislature may deem requisite to carry out the provisions of this section.[2]

The second California commission appointed in 1917 was unanimously in favor of compulsory health insurance. In November 1918, the people of California in a referendum vote rejected the amendment to the constitution designed to give the legislature power to pass social insurance legisla-

[1] *Laws of 1915*, ch. 275.

[2] *California Social Insurance Commission Report*, Jan., 1917, p. 17.

tion.[1]   Prof. John R. Commons, in commenting on the large vote against the amendment, stated as follows: " In addition to the opposition of Christian Scientists, the fraternal and commercial insurance companies assisted in defeating the amendment.   Some of the large fraternal organizations sent a personal communication to each of their members asking that they vote against the amendment." [2]   Following the popular rejection of the constitutional amendment, the re-created social insurance commission submitted a report in March 1919, which contains, in outline form, a bill with certain standards agreed upon as essential to any bill which would be adaptable to California conditions.[3]

### CONNECTICUT

The Connecticut commission appointed in 1917 to study, among other subjects, health insurance, reported in January 1919.   The findings of the commission were definitely adverse to the establishment of a compulsory system, stating: " In our opinion, the General Assembly may, with entire propriety, postpone further legislative consideration of this phase of social insurance until the changes in our national, state and personal relations resulting from the war have been fully readjusted." [4]

### ILLINOIS

The health insurance commission created by the legislature in 1917, made a voluminous report in May, 1919.   The commission recognized the seriousness of the sickness problem, and the justification of applying the insurance principle to the sickness hazard,[5] but the majority report was adverse

[1] *Illinois Health Insurance Commission Report*, p. 646.

[2] *Ibid.*, pp. 646-647.

[3] *California Social Insurance Commission Report*, 1919, pp. 10-12.

[4] *Connecticut Commission on Public Welfare Report*, 1919, p. 16.

[5] *Illinois Health Insurance Commission Report*, pp. 163-164.

to the establishment of a compulsory health insurance system. The findings were that about 30 per cent of the wage-earners of the state had some type of health insurance, provided by one or more of the following agencies—unions, establishment funds of different kinds, fraternal and benevolent societies, independent foreign societies, mutual and stock companies.[1] The commission admitted that voluntary health insurance in the state was not standardized, but expected such insurance to show both a growth and tendency toward standardization as its importance received more general recognition. Particular objection was taken against the idea of compulsion. Certain recommendations were made relating to the improvement of public health work, including tuberculosis, venereal disease, study of the problem of compensation for occupational diseases, study of infant and maternal mortality, and public health nursing development on a county basis.[2]

The minority report, signed by Dr. Alice Hamilton and another member of the commission, dissented vigorously from the majority report. The minority stated as follows:

. . . We feel that the recommendations made fall far short of what is called for by the facts disclosed by the Commission's investigations. Moreover, we do not agree with the majority of the Commission in their conclusions relative to compulsory health insurance. We believe that the results of the investigations made for the Commission are conclusive evidence of the need for a system of compulsory health insurance which would be applicable to practically all members of the wage-earning group, would more equitably distribute the burden of the costs of sickness and would make more adequate provision for the medical care of wage-earners and their dependents who become sick. . . . We believe that the conclusion of the majority

[1] *Ibid.*, p. 164.
[2] *Ibid.*, pp. 167-168.

of the Commission is founded on an inadequate and somewhat misleading interpretation of the facts disclosed by the investigations made for the Commission.

The propositions advanced in the minority report were that: (1) sickness is a serious problem; (2) sickness is a problem calling for the application of the insurance prin‑ciple; (3) it is generally recognized that sickness is an insurable risk; (4) in spite of the fact that there are many varieties of carriers in the state, the great majority of wage-earners have no health insurance; (5) much of the health insurance carried by wage-earners is inadequate and costly; (6) most of the health insurance carried provides for partial indemnity for lost wages only; (7) there is a distinct need for a better organization of medical service for wage-earners; (8) if the application of the insurance principle to the problem of sickness among wage-earners and their dependents is to be most effective, it must be universal.[1]

### MASSACHUSETTS

The first commission to report (February 1917) was that created in 1916 by the General Court of Massachusetts. This commission investigated sickness insurance, old age pensions, unemployment and hours of labor. The commission unanimously indorsed the principle of health insurance,[2] five members of the nine on the commission favoring the immediate adoption of compulsory health insurance; two that there should be more investigation, and two that if a compulsory system were adopted employees should not be required to contribute.[3]

[1] *Ibid.*, pp. 169-173.

[2] *Massachusetts Special Commission on Social Insurance Report*, Feb., 1917.

[3] National Industrial Conference Board, *Research Rept.*, no. 6, May, 1918, p. 19.

The second Massachusetts commission continued the study begun by the first, concentrating on health insurance. It reported January 15, 1918. Of the eleven members, nine voted against compulsory health insurance, the remaining two voted [1] in favor without contributions from the employee.[4] Prof. John R. Commons, in commenting upon this report stated as follows: " From a perusal of the report it would seem that the commission was concerned particularly with the problem of ascertaining the attitude of different interests. The report states that an analysis of the evidence reveals no growing demand in the commonwealth for compulsory contributory health legislation." [2]

### NEW JERSEY

A commission was appointed in New Jersey in 1911 on old age insurance and pensions.[3] This commission thought health insurance of more pressing importance than old age pensions, and decided that " health protection should precede any provision for old age pensions." The same commission, re-appointed in 1917 reported as follows: [4]

The Commission is of the opinion that health insurance is a measure which gives great promise both of relieving economic distress due to sickness and of stimulating preventive action. To achieve these ends, such a measure, adapted to New Jersey's needs, should be based upon the following fundamental principles. Existing health insurance agencies that are conducted on an adequate basis at actual cost should, with mutual man-

[1] *Report of the Special Commission on Health Insurance of Massachusetts.* For majority report see pp. 36-55; for minority report, pp. 61-70.

[2] *Illinois Health Insurance Commission Report*, p. 630.

[3] *Laws of 1911*, ch. 198.

[4] *Report on Health Insurance by the New Jersey Commission on Old Age, Insurance and Pensions*, Nov., 1917, p. 2.

agement, be utilized in the further development of a comprehensive health insurance system. In order that the greater effectiveness and economy of a universal system may be enjoyed, health insurance should be made to cover all regularly employed wage-earners. Insurance should provide medical care and health instruction in order that its work may be both curative and preventive. To minimize the financial distress attending sickness the system should provide a cash benefit during temporary incapacity to work. It should also provide maternity care to meet the special needs of working mothers. Health insurance should be democratically supported and managed by those directly concerned, the state bearing as its share the cost of general administration as it does in workmen's compensation.[1]

### OHIO

The Ohio Health and Old Age Insurance Commission was composed of seven members, exclusive of a secretary and a director of investigations, both of the latter not voting on the findings of the commission. In the majority report the commission made a number of recommendations pertaining to the organization of public health work in general. In addition, the following were the recommendations pertaining to health insurance: (1) the principle of health insurance is approved as a means of distributing the cost of sickness; (2) health insurance should be required for all employees to be paid for by employers and employees in equal proportion. The state should pay all costs of state administration as in the case of the workmen's compensation act, and all costs of supervision of insurance carriers; (3) the benefits to workers under health insurance should consist of cash payment of a part of the wages of workers disabled by sickness; complete medical care for the worker, including hospital and home care and all surgical attendance and the cost of all medicines and appliances; adequate provision for

[1] *Ibid.*, pp. 19-20.

rehabilitation, both physical and vocational, in cooperation with existing public departments and institutions; dental care; medical care for the wives and dependents of the workers if the same can be done constitutionally; and a burial benefit for the worker; (4) the exact form of organization of the medical service, including hospital and dental service, should be left largely to the State Health Insurance Commission, which administers the act, to develop plans to meet conditions in different parts of the state; it should be clearly established that medical, hospital and dental care shall be adequately compensated; (5) the insurance should be carried in establishment funds mutually managed and in public mutual associations.   Companies or associations writing insurance for profit shall not be permitted to be carriers of such insurance; (6) the system should be administered by a state health insurance commission.[1]

The dissenting opinion held that the compulsory feature was wrong in principle for it would mean the sacrifice of the independence of a large number of people; the proposed distribution of the cost of 50 per cent by employer and 50 per cent by employee did not equitably place the burden upon society as a whole; sufficient study was not made of the entire question to warrant its early adoption in Ohio.[2]

### PENNSYLVANIA

The Pennsylvania Health Insurance Commission, appointed under the legislative act of July 25, 1917 was directed to investigate among other things: (1) sickness and accident of employees and their families, not compensated under the workmen's compensation law; (2) the adequacy of the present methods of treatment and care of such sick-

[1] *Report on Health Insurance and Old Age Pensions*, by the Ohio Health Insurance and Old Age Commission, Feb., 1919, pp. 16-18.

[2] *Ibid.*, pp. 19-20.

ness and injury; and (3) the adequacy of the present methods of meeting the losses caused by such sickness or injury, either by mutual or stock insurance companies or associations, by fraternal or other mutual benefit associations, by employers and employees jointly, by employees alone or otherwise.[1] The investigations carried on, limited though they were, indicated that sickness among wage-earners in Pennsylvania presented a problem of vital importance not only to employees, but to industry and to the state as well. The committee recommended a careful study of possible remedial legislation covering: (1) such adequate medical care for employees and their families during sickness as will materially improve health conditions; (2) a means of meeting the wage-loss ordinarily suffered by employees during periods of illness by methods which will apportion fairly the economic burden between the three elements responsible—the individual, industry and the state; (3) a special study of proposed and existing systems of health insurance in this and other countries.[2]

The enactment by the general assembly of 1919 of an act creating a commission for the study of health insurance was the direct result of the report to that assembly of the health insurance commission appointed under the act of July 25, 1917. The first commission had an appropriation of only $5,000 to carry on its work. The second commission spent but little time in investigation and the majority based its report largely on testimony given at various public hearings, and brought in an adverse report. The minority report stated that the commission had really not made a careful study of health insurance; that the commission was not therefore in a position to know what a careful study of

[1] *Report of Health Insurance Commission of Pennsylvania*, Jan., 1919, p. 1.

[2] *Ibid.*, p. 36.

health insurance might disclose; and that nothing happened in the preceding two years could change the validity of the recommendation of the first health insurance commission that the economic burden of sickness should not be allowed to fall on the sick wage worker.[1]

### WISCONSIN

The Wisconsin committee, composed of two state senators and three assemblymen, held its first meeting in Milwaukee on September 12, 1917. In order to determine upon the proper line of procedure a general public meeting was called on October 9, 1917. This meeting was widely advertised and various organizations interested in the subject matter were represented. The representatives present were requested to suggest to the committee what scope the investigation should take, and the best method to obtain information on the subject. While the legislative resolution and act authorizing the appointment of the committee directed a study of the whole subject of social insurance, the committee after this meeting, decided to limit itself to a study of the subject of health insurance.[2]

The majority report, signed by four of the five members, stated that before adopting legislation of this nature, an effort should be made first to extend and make more efficient the existing preventive agencies. The committee further stated that certain preventive measures should be undertaken, such as making more liberal appropriations to the State Board of Health, more widespread medical examinations in the public schools, establishment of district nursing centers, etc. Finally, the recommendation was made that

[1] Appendix to the *Pennsylvania Legislative Journal*, Apr. 18, 1921, pp. 3925-3926.

[2] *Report of the Special Committee on Social Insurance of Wisconsin*, Jan. 1, 1919, p. 9.

the State Insurance Department concern itself more actively with the supervision and control of voluntary insurance undertakings, having for their object the pecuniary relief or medical attendance, or both, of wage-earners during more or less prolonged periods of illness. It was felt that a standardized plan of organization and procedure might be worked out under the direction of the insurance commissioner and recommended to the wage-earners of the state as well as to the employers of labor for individual or collective adoption.[1]

The minority report, signed by one member, largely disagreed with the majority contention regarding the need of a constitutional amendment before health insurance could be enacted, and recommended that a compulsory health insurance law be enacted, embodying the general scheme of the " Standard Bill " of the American Association for Labor Legislation.[2]

### SUMMARY

Agitation for health insurance legislation in the several states above mentioned was followed by the introduction of bills proposing health insurance in some of the states. The Nicoll bill, introduced in the New York legislature in February 1918, was a modification of the " Standard Bill " and received considerable backing. However, thus far, nothing tangible has come out of these early efforts for health insurance in this country.[3]

---

[1] *Ibid.*, pp. 49-51.

[2] *Ibid.*, pp. 51-52.

[3] A health insurance bill was actually passed in the New York Senate. To forestall affirmative action by the Assembly, it has been maintained by some that the opponents of the bill presented an occupational disease compensation bill, which was finally passed instead of the health insurance bill. See: *U. S. Bur. Lab. Stat. Mo. Lab. Rev.*, vol. vi, no. 5, May, 1918, pp. 227-230. See also: *Amer. Labor Legis. Rev.*, vol. xv, 1925, p. 317.

### ARGUMENTS RAISED AGAINST HEALTH INSURANCE

Many arguments have been presented against the adoption of health insurance, and particularly of compulsory health insurance in this country. The plan has been characterized as un-American, Socialistic, etc.,[1] while other more critical reasons have been advanced against its adoption. Following are some of the arguments which have been presented against the plan, and which permeate most of the discussions of those who oppose health insurance in this country.

Compulsory health insurance has been held undesirable because, while it is admitted that sickness is a serious problem in the United States, the mortality and morbidity experience is more favorable than that of Germany and Austria, which have compulsory health insurance,[2] and the problem is one of sickness prevention rather than sickness insurance;[3] a large percentage of the cases of poverty caused or accompanied by sickness would not be avoided by compulsory health insurance;[4] the burden proposed to be placed upon the state and all employers for part payment of the cost of insurance for all diseases of all wage-earners would compel the state and employers to pay for that which they did not cause and for which they are not responsible in any real or tangible sense;[5] the insurance principle applies to sickness not

An old age, sickness and disability insurance bill, along the lines of the earlier "Standard Bill" and applying only to such cases as are not already covered by the Workmen's Compensation Law, was introduced in the New York Assembly in Jan., 1931. (Assembly Intro., no. 5.)

[1] Hoffman, F. L., In *Report 17th Annual Meeting National Civic Federation*, Feb. 3, 1917, pp. 165-166.

[2] Hoffman, F. L., *Facts and Fallacies of Compulsory Health Insurance* (Newark, 1917), p. 89.

[3] National Industrial Conference Board, *Research Report*, no. 6, (New York, May, 1918).

[4] *Illinois Health Insurance Commission Report*, p. 165.

[5] *Ibid.*, p. 165.

only through the compulsory insurance systems of Europe, but also through the voluntary insurance available in the United States;[1] existing insurance agencies are adequate and more efficient than a state-owned or controlled system;[2] the proposal for proportional contributions by employer, employee and the state is based solely upon expediency;[3] there is no evidence that compulsory insurance has resulted in an improvement in health;[4] compulsory health insurance would involve such large sums to be carried in state or local funds that political control and management would result;[5] the status of the medical profession would be lowered because of overwork, burdensome details, suppression of initiative on the part of physicians, and insufficient remuneration.[6]

### SUMMARY

Detailed answers to the above arguments have been presented in some of the majority and minority reports mentioned above,[7] and more especially by the extensive brief of the American Association for Labor Legislation.[8]

Some have indicated that the principal opposition to compulsory health insurance comes from so-called burial insurance companies, or those writing a large number of industrial insurance policies, and also from the casualty insurance

[1] *Ibid.*, p. 633.

[2] *Ibid.*, p. 633.

[3] *Ibid.*, p. 165.

[4] National Industrial Conference Board, *Research Report*, no. 6, p. 18.

[5] *Illinois Health Insurance Commission Report*, pp. 165-166.

[6] *Ibid.*, p. 633.

[7] *Ibid., Minority Report*, pp. 171-172, 631-633; *Report on Health, Health Insurance, Old Age Pensions*, by Ohio Health Insurance and Old Age Commission, pp. 97-113.

[8] *American Labor Legis. Rev.*, vol. vi, 1916, pp. 155-236.

companies.[1]  It is apparent, from the experiences in Germany and England, as well as in other European countries, that health insurance under governmental auspices has become a definite part of the lives and policies of the people of these lands.  Arguments similar to those presented in this country were propounded in countries where health insurance is now in effect.  By and large, sickness experiences of comparable industrial populations are about the same and health measures taken with success for one group should generally be applicable to other working groups.

One question that may be raised is the relation of a proposed system of health insurance to existing workmen's compensation laws covering accidents and occupational diseases.  This matter came up in England when health insurance was introduced.  Not only has the health insurance legislation not relieved the employers of any duties under the compensation system, but it is specifically provided that when a disabled workman receives his compensation benefits he is not to draw any sick benefits.  Also, it has been provided that the weekly sickness or disablement benefit is only payable insofar as it exceeds in amount the weekly value of the workmen's compensation payment.  If the insured person refuses to comply or sue for compensation, the sickness insurance institution may itself take action in his behalf.  Conditions are quite similar in this country to those in England, and it may be fair to assume that no serious conflicts would arise between the enforcement of existing compensation laws and a system of health insurance for occupational diseases.

Health insurance has been utilized in various industrial countries to solve problems similar to those which face this country at the present time.  The foundations for health insurance have already been laid through experiences of the

[1] Lapp, John A., *Health Insurance*, National Conference of Charities and Correction.  Report 1919, pp. 442-447.

commercial and mutual insurance companies, and various fraternal and labor organization sickness funds, as well as other less accurately controlled sick and benevolent associations.

### CONSTITUTIONALITY OF HEALTH INSURANCE

The experience with the enactment of the first workmen's compensation law, mention of which has already been made, leads one to believe that any first attempt at the enactment of a health insurance plan in this country would be challenged as to its constitutionality. Since there has not as yet been enacted any health insurance bill, the occasion has therefore not arisen seriously to consider the question of constitutionality. However, no discussion of this subject would be complete without at least a passing mention of this matter.

In legislation of this kind, the provisions of the 14th Amendment would presumably be brought into discussion. It has been pointed out by competent authority that the first limitation upon the legislature on the constitutionality of such legislation is contained in the 14th Amendment to the Constitution of the United States and the similar provisions found in nearly all the state constitutions: " nor shall any state deprive any person of life, liberty, or property without due process of law, nor deny to any person within its jurisdiction the equal protection of the laws." The first prohibition brings before the United States Supreme Court any legislation affecting individual liberty and property; the second prohibits arbitrary and unreasonable action in regard to individuals or classes of individuals, so that a compulsory health insurance law would have to be submitted to judicial scrutiny for final settlement.[1]

Prof. Freund, in discussing this question, noted that the

[1] Chamberlain, Joseph P., *Constitutionality of Health Insurance*, in Rubinow, I. M., *Standards of Health Insurance*, 1916, p. 275.

objection to such a system would be that an individual would be forced to share in making good a loss with which in a particular case he had no connection, although he took the utmost precaution to avoid such loss so far as the management of his own property was concerned. The controlling consideration is the existence of a risk or danger, which the police power may seek to minimize; and it is reasonable that those who create or maintain the risk or danger for their own benefit should consent to the most effectual means of obviating its harmful consequences. Collective responsibility is a wise and conservative method of meeting the risk, and its imposition should be allowed as a valid condition of the right of keeping a dangerous instrument.[1]

Others who have maintained that health insurance would be found to be constitutional, have held that the enactment of health insurance legislation would be a proper exercise of the police power of the state, and even though it is impossible adequately to define either " police power " or " due process ", the hope of clarifying the ideas which they represent is by critical examination of the conditions by which they are brought into operation. While the jurists of early American history were imbued with the inviolability of individual rights, the right and power of the state to impose reasonable restrictions has gained gradual judicial recognition.[2] Compulsory health insurance has further been held to be a public purpose, and is so reasonable a method of reducing the risk of pauperism, and the necessity of other forms of public relief, that it ought to be regarded as a legitimate exercise of the police power. The only way to settle the question would be to legislate and to leave the decision to the courts.[3]

[1] Freund, E., *The Police Power* (Chicago, 1914), p. 460.

[2] Harper, F. V., "Due Process of Law in State Labor Legislation," *Michigan Law Review*, April, 1928, p. 599.

[3] Ellingwood, A. R. and Coombs, W., *The Government and Labor* (New York, 1926), p. 615.

## BRIEF FOR HEALTH INSURANCE

In the brief for health insurance prepared by the committee of the American Association for Labor Legislation, and out of which brief was developed the so-called "Standard Bill" to which attention was previously drawn, the following six points were emphasized: [1]

1. High sickness and death rates are prevalent among American wage-earners.
2. More extended provision for medical care among wage-earners is necessary.
3. More effective methods are needed for meeting the wage loss due to illness.
4. Additional efforts to prevent sickness are necessary.
5. Existing agencies cannot meet these needs.
6. Compulsory contributory health insurance providing medical and cash benefits is an appropriate method of securing the results desired.

This brief was carefully drawn up by a group of experts and fully documented to sustain the six points mentioned above. It is deemed unnecessary to attempt a detailed discussion of these various subjects, other than to mention that the contentions raised in these points are held to be as valid at present as they were in 1916 when first drafted.

The health insurance standards which the above-mentioned committee drew up as a further guide in the drafting of the "Standard Bill" are indicative of the methods to be followed, perhaps with certain adjustments, in attempting to draft and enact legislation at the present time. These health insurance standards are as follows: [2]

[1] *American Labor Legislation Rev.*, vol. vi, 1916, p. 155. For full discussion see pp. 155-236.

[2] *Ibid.*, pp. 237-238.

1. To be effective health insurance should be compulsory, on the basis of joint contributions of employer, employee and the state.
2. The compulsory insurance should include all wage-earners earning less than a given annual sum, where employed in sufficient regularity to make it practicable to compute and collect assessments. Casual and home workers, should, as far as practicable, be included within the plan and scope of a compulsory system.
3. There should be a voluntary supplementary system for groups of persons (wage workers or others) who for practical reasons are kept out of the compulsory system.
4. Health insurance should provide for a specified period only, provisionally set at twenty-six weeks (one-half year), but a system of invalidity insurance should be combined with health insurance so that all disability due to disease will be taken care of in one law, although the funds should be separate.
5. Health insurance on the compulsory plan should be carried by mutual local funds jointly managed by employers and employees under public supervision. In large cities such locals may be organized by trades with a federated bureau for the medical relief. Establishment funds and existing mutual sick funds may be permitted to carry the insurance where their existence does not injure the local funds, but they must be under strict government supervision.
6. Invalidity insurance should be carried by funds covering a larger geographical area comprising the districts of a number of local health insurance funds. The administration of the invalidity fund should be intimately associated with that of the local health funds and on a representative basis.
7. Both health and invalidity insurance should include medical service, supplies, necessary nursing and hospital care. Such provision should be thoroughly adequate, but its organization may be left to the local societies under strict government control.
8. Cash benefits should be provided by both invalidity and

health insurance for the insured or his dependents during such disability.

9. It is highly desirable that prevention should be emphasized so that the introduction of a compulsory health and invalidity insurance system shall lead to a campaign of health conservation similar to the safety movement resulting from workmen's compensation.

In the tentative " Standard Bill " which was drafted the various phases of a health insurance plan are included, such as persons insured; compulsory insurance to cover all earning less than \$100 a month; benefits, with cases in which paid; minimum benefits; beginning of right; medical, surgical and nursing attendance; medical service through panel system; medical and surgical supplies; hospital treatment; cash benefit ( ⅔ of weekly wages) ; cash benefit to dependents; periods of payment; maternity benefits; additional benefits; contributions; insurance carriers, as (1) state fund managed entirely by the state, (2) by approved societies, as in England, and (3) by district mutual associations, as in Germany; voluntary insurance; approved societies; employers' contributions; state contributions; supervision through state social insurance commission; social insurance council; medical advisory board; and settlement of medical and other disputes.[1]

The preceding summary is perhaps a sufficiently complete outline of the functions of a health insurance plan and the type of organization which would be required to put such a plan into effect. It is recognized that the various stipulations are each and every one of them probably open to question and discussion, though they represent the fundamentals

[1] *Ibid.*, pp. 239-268. Regarding state funds managed entirely by the state, see: *New York Workmen's Compensation Law*, art. v, para. 90-91, stating in regard to such funds: " Such fund shall be administered by the industrial commissioner ", and " The state treasurer is custodian of the state insurance fund."

about which such a compulsory health insurance plan would have to be built.[1]

All persons are exposed to the risk of sickness, and its occurrence entails for them or their dependents either an increase or expenditure for medical treatment and drugs, and a reduction or even cessation of income resulting from incapacity for work. Usually both an increase in expenditures and a limitation of income occur. Health insurance has for its object the mitigation of the economic consequences of sickness. In its thorough application, it provides for these emergencies among the wage-earners. It is usually considered unnecessary to impose compulsion to insure against contingencies of sickness upon persons who possess sufficient means to enable them to make their own arrangements for insurance, as the practice of voluntary insurance is widely developed among the richer classes. In some countries recent legislation has gone beyond the limits of the wage-earning class, and the essential criterion for the determination of the scope of compulsory health insurance has become the annual income of the individual. In this way the compulsory principle has been extended to all persons of small means.[2]

[1] For some objections to the " Standard Bill," see: *U. S. Bur. Lab. Stat. Mo. Lab. Rev.*, vol. iv, no. 4, April, 1917, pp. 501-502. It is contended that unions, fraternal societies, and other voluntary organizations engaged in health insurance should be given a place under a state system and should be encouraged to continue to develop their activities; the compulsory system should not drive out the voluntary; administration should not be left in the hands of employers and employees through district mutual associations because of the danger of deadlocks on disputed issues; the " Standard Bill " does not provide for the selection of the best type of executives; cash and medical benefits should be entirely separated. See also: Sherman, P. T., *Criticism of a Tentative Draft of an Act for Health Insurance* (New York), 1917, 93 pp.

[2] Inter. Labour Office, Series M, no. 6, *Compulsory Sickness Insurance*, 1927, pp. 29-30.

The motives which have prompted compulsion have rested upon the desire to make certain that all elements in the population most in need of protection should be included in the insurance scheme. It was for this reason that the health insurance plan of Germany met with initial success.[1] While compulsory health insurance has usually been applied only for wage-earners, there has not always been a definition of the term "wage-earner." The presumption has been that those responsible for the enforcement of the laws were fully apprised of the exact meaning of the term, and that in doubtful cases the courts could be appealed to for an exact definition. It has generally been presumed that in order to be liable to insurance a person must be engaged in: (1) work in a dependent position; (2) work under a contract; and (3) work as the ordinary means of livelihood.[2] Through the application of this definition most of those who were presumably in need of such protection and service as health insurance provided, would be covered.

Even though compulsion in general is somewhat objectionable to Americans, it cannot in any sense be as objectionable as the suffering which results from illness or improvidence. The principle of compulsion is now used to enforce education, sanitation, fire prevention, food supervision, etc., as well as in workmen's compensation. Most important of all, the careless, the indifferent, the impoverished— the very ones who need the protection offered by a system of compulsory health insurance, are those who would ordinarily fail to avail themselves of some type of voluntary insurance.

Those of low income must in some way be protected against the hazards of occupational and other diseases and

[1] Frankel, Lee K., "Workmen's Insurance Compensation," *U. S. Bur. Lab. Stat. Bull. no. 212*, p. 599.

[2] Inter. Labour Office, Series M, no. 6, p. 31.

ailments. The industries in which they are engaged should, from a social point of view, help to bear the burden of their sickness and loss of wages, and to initiate preventive measures following a careful study of sickness experience. Through compulsory health insurance these objects can be achieved.

### EXTENT OF SICKNESS IN THE UNITED STATES

Professional workers of relief agencies have long recognized that sickness represents the most frequent factor of individual destitution. The growth of the various health and life conservation movements, the organization and development of the anti-tuberculosis movement, the great increase in diseases of middle life and the high mortality caused by these diseases—all have brought to light not only the social waste caused by excessive and preventable illness, but also the economic conditions which are responsible for them.[1]

Various studies have attempted to compute the average number of days lost by workers during a year as a result of sickness alone. The United States Commission on Industrial Relations, after investigating representative establishments and occupations involving nearly a million workers in this country, estimated that nine days a year was the average loss per worker due to sickness.[2] These figures do not take into consideration the effects of sickness of the worker upon his vigor and efficiency, or upon the welfare of his family, or upon the efficiency of the industry in which he is engaged. Further, the figures do not indicate the losses due to death, and only inadequately afford a true conception of the problem of sickness among wage-earners.

[1] Rubinow, I. M., *Standards of Health Insurance* (New York, 1916), p. 6.

[2] *U. S. Public Health Bull. no. 76,* Mar., 1916, p. 6.

A number of other careful studies have been made of the incidence of sickness among the working population. Brundage analyzed the causes of absence from work of 6,700 employees in one manufacturing company. He found that absence, including that for sickness of more than two days' duration, amounted to 17.1 working days per year, of which 5.4 days were due to cases of sickness that disabled for more than two consecutive working days.[1] In another study of 1,282 office workers, largely women, the average morbidity was 8.15 days per person per year.[2] A study covering five years sickness experience in Boston, indicated that in the particular company sickness caused 20 times as many cases of absenteeism as accidents, and was responsible for seven times as much loss of time from work.[3] In a careful study of the incidence of illness in a general population group in Hagerstown, Md. from 1921-1924, it was found that over 100 cases of illness occurred annually for each death, and that more than half of the morbidity was due to respiratory diseases.[4]

A most interesting analysis of the problem of sickness was made by Homer Folks in an address entitled *The Distribution of the Costs of Sickness in the United States*. In this study Folks pointed out the various elements that annually enter into the sickness bill in the United States, including the

[1] Brundage, D. K., "Sickness and Absenteeism During 1919 in a Large Industrial Establishment," *U. S. Public Health Rept.*, Sept. 10, 1920, 14 pp.

[2] Brundage, D. K., "Sickness among Office Workers," *U. S. Public Health Rept.*, Mar. 10, 1922, 7 pp.; "Sickness among a Group of Industrial Employees," *U. S. Public Health Rept.*, Jan. 17, 1930.

[3] Sappington, C. O., "A Five Years' Sickness and Accident Experience," *Jo. Ind. Hyg.*, vol. vi, no. 3, July, 1924, p. 94.

[4] Sydenstricker, E., "The Incidence of Illness in a General Population Group," *U. S. Public Health Rept.*, vol. 40, no. 7, Feb. 13, 1925, pp. 279-291.

expenditures for physicians, quacks, dispensaries, hospital care, nursing care in patients' homes, medicines, medical supplies, dental care, loss of wages during illness, loss of future net earnings on account of premature death due to illness and preventive health work. He estimated that the total annual cost of illness to the people of the United States was not less than the enormous sum of $15,730,000,000.[1] This figure does not include the reduced earnings due to slight illness. Folks comments upon the fact that there is no satisfactory explanation of the fact that the various American states have within recent years provided very extensively for the distribution of losses arising from accidents, but have made no similar provision as to losses arising from illness.[2]

The realization of the widespread nature of illness has recently resulted in the organization of the Committee on the Costs of Medical Care. This Committee has already issued some preliminary studies, indicating, among other things, that the people of the United States, on the average, suffer from one to two disabling illnesses per year,[3] and that there are approximately 1,500,000 persons whose services are utilized in this country in one way or another in the care and prevention of illness.[4] These figures are indeed indicative of the seriousness of the problem of sickness as affecting the entire population of the country.

[1] Folks, Homer, *The Distribution of the Cost of Sickness in the United States*, address in Paris, July 12, 1928, p. 25.

[2] *Ibid.*, p. 36.

[3] Committee on the Costs of Medical Care, *Publication no. 2* (Washington, 1929), p. 5.

[4] *Ibid.*, p. 10.

## MORTALITY OF THE INDUSTRIAL POPULATION

In the preceding discussion of health insurance it is quite apparent that there has been no particular stressing of the application of health insurance solely to instances of occupational diseases, but rather to the general application of the principle to all cases of illness among the wage-earners of this country. The question might possibly arise as to the validity of the thesis that compulsory health insurance should be applied to the problem of occupational diseases, when, at the same time, the matter of sickness is so widespread that if health insurance is a desirable measure, it should be provided for all wage-earners. As has previously been indicated, the present machinery for caring for cases of occupational diseases through workmen's compensation is inadequate to meet the problem. Further, the possibility of reaching all, or most of the cases of occupational diseases through methods now in force or likely to be extended or developed in the near future, is not probable. This is primarily due to the fact that occupational diseases, broadly conceived, are more widespread than is generally assumed and that the health and lives of the workers in industry are more adversely affected than has heretofore been recognized.

It is due to the work of Dr. Louis I. Dublin that valuable information on this subject has recently become available.[1] Dr. Dublin prepared a report on the mortality experience among three and one-quarter million white male wage-earners insured in the Industrial Department of the Metropolitan Life Insurance Company. The study covered the three-year period 1922 to 1924 inclusive, and involved a study of work-

[1] Dublin, Louis I., " The Mortality Trend in the Industrial Population," *Amer. Jo. Public Health*, vol. xix, no. 5, May, 1929, pp. 475-481. This indicates the difficulties involved in attempting to compensate for so-called occupational diseases. Pneumonia, ordinarily not included as an occupational disease, appears to be about 50 per cent higher among workers in the iron and steel industries than among other industrial employees.

ers in the United States (and some in Canada) who earn their livelihood in the manufacturing plants, mines, transportation industries and other mechanical pursuits. They constitute a fairly homogeneous social and economic group which may be described as the urban wage-earning population.

The adult males of this class were found to have a higher mortality and diminished longevity than those in other forms of employment. The mortality rate for the group of insured males was found to be higher than that of all males living in the United States registration states in every age period from 25 to 64 years, the period of industrial employment. In the registration states, the death rate per 100,000 for all causes of death among white males in 1923, 25 to 34 years, was 457.7; for Metropolitan Industrial Department 556.1; ages 35 to 44, 717.4 and 946.6; 45 to 54, 1,207.4 and 1,725.4; 55 to 64, 2,527.8 and 3,385.3.

These rates reflect the results of industrial exposure. In the beginning, that is between the ages 15 to 24, the group of policyholders is in its best physical condition and has a lower mortality rate than the general population. At age 25, the situation changes; and thereafter, as Dr. Dublin indicates, largely as a result of industrial hazards, the mortality rate is heavier than that of the general male population.

The effects of industry are also clearly revealed in the differential death rates for males and females of the same economic class. After age 10, the mortality rates for males was found to be consistently higher than for females, with the single exception of the age period 15 to 24 years. Since the industrialization of women is then at its height, the question of the relative susceptibility of men and women to the exactions of industry upon their health is definitely raised. After this age, the large majority of these women become housewives, and a lower rate of mortality in each age group than for men becomes evident.

The comparison between the death rates of industrial males and others is even more marked. Among the industrial males, 20 years of age and over, the death rate per 100,000 for all causes of death in 1923 was 1,404.1, whereas for so-called ordinary male policyholders in the same company the figure was only 500.0. In terms of life expectation, the severer status of the industrial worker is equally impressive. The industrial worker at age 20, when he begins his career, has an expectation of life of 42 years; or, in other words, he may expect to reach 62. The 20-year-old worker engaged in non-hazardous occupations, however, may expect to obtain age 69, or seven years additional.

The above comparisons between the death rates of various groups, has been held by Dr. Dublin to show clearly the influence of industrial environment on mortality rates and life expectation. The difference in the mortality rates for these groups gives a rough measure of the tax which industrial work exacts, and reflects the hazards to which workers are exposed. Other items may obviously account for a part of the disparity. Heredity and innate differences play some part, but probably the most important factors are the conditions incidental to industrial employment including deleterious dusts, excessive fatigue, bad posture, crowded workrooms, dampness, extreme changes of temperature and sometimes specific occupational poisoning, to which industrial workers are frequently exposed.[1]

### CONCLUSIONS

In various discussions regarding the advisability of permitting the individual worker to take care of his sickness insurance requirements, the question as to whether or not the worker is able to do so unaided financially is usually passed over. Some interesting figures bearing on this question are

[1] *Ibid.*, p. 478.

available, particularly insofar as wages and cost of living are concerned. For instance, the United States Census of Manufactures Report [1] indicates that in 1925 the average number of factory wage-earners in New York City was 538,845, and that the wages paid to these workers in 1925 totaled $844,648,136.00, making an average annual wage for this large group of workers of $1,567.00. The New York State Department of Labor, in figures published in September 1928, showed that during August 1928, a total of 147,613 male and female employees in the state averaged a weekly wage of $31.92, with the males averaging $37.59 or at the rate of $1,954.68 per annum, and the female workers $21.24 or at the rate of $1,104.48 per annum. This computation of annual earnings does not make any allowance for seasonal fluctuation in work, time lost for other reasons as sickness, breakdown of machinery, etc.[2] The National Industrial Conference Board made a study of wages in the United States, and reported that in 1928 the average weekly earnings of wage-earners in 25 industries equalled about $27.00, or on the basis of about $1404.00 per annum if full time were computed.[3] Other studies covering large numbers of workers have similarly indicated the low average annual income of industrial workers.[4]

On the other hand, analyses of the cost of living have generally agreed on the point that if minimum allowance were to be made for life and sickness insurance, medical care for the worker and his family, and other essentials of a workingman's family including himself, his wife and two

[1] *U. S. Census Report on Manufactures*, 1927.

[2] N. Y. State Dept. of Labor, *Industrial Bull.*, Sept., 1928.

[3] National Industrial Conference Board, *Wages in United States 1928* (New York, 1929), pp. 5-10.

[4] National Bureau of Economic Research, *Income in the United States 1909-1919*, vol. ii (New York, 1922).

children, the worker would need much more income than he now receives.[1] In view of the difference between income and the cost of necessities, such as food, rent, etc., the worker frequently finds it impossible to make proper provision for sickness through insurance or otherwise.

The reasons for the need of sickness or compulsory health insurance have been aptly summarized as follows:[2] (1) Sickness is essentially a public matter. The health, length of life, sense of security against overwhelming disaster and freedom from pauperization of workmen concern the entire community. (2) By no other means can more than a small number be reached. The high commissions in the voluntary plan involve prohibitory cost for many; lower commissions might secure even a smaller number of insured persons. (3) Those who are not reached by voluntary insurance— the careless, the indifferent, the impoverished, are precisely those who need the protection most and whose care is important for the community welfare, in order to prevent the spread of disease and to improve the longevity and efficiency of workmen. (4) By this means only can a coherent, consistent, fairly-administered system be operated, bringing the benefits, both cash and medical, to all who should have them. (5) It is much more economical, as well as efficient in operation and gives maximum benefits at minimum cost. (6) If not compulsory, the workman's insurance binds him to his job while the unenlightened employer who refuses to contribute seems to have an advantage; this is not well for the employer nor for the employee. The in-

---

[1] National Industrial Conference Board estimated $1,659.84 was needed in a family with two children. See: Research Reports 22, 24; Special Reports 7, 8, 13, *U. S. Dept. Lab. Stat. Mo. Lab. Rev.*, Dec., 1919: " Tentative Quantity Cost Budget Necessary to Maintain Family of Five in Washington, D. C.," estimated $2,288.25 needed with three children.

[2] Dawson, Miles M., " Why Compulsory Insurance? ", *Amer. Labor Legis. Rev.*, vol. vii, 1917, pp. 117-118.

surance should follow the employee as long as he keeps within the domicile and all employers should contribute ratably. (7) Upon no basis, other than compulsion, can the chief social purposes be served or the state's contribution be justified. The state desires well-being, care when disabled, means for prompt and effective care, prevention of disease, prevention of pauperization for all its workmen and their families. By this means these benefits can be secured; and, under such a public plan the state would be justified in imposing a special charge upon employers and employees and in itself contributing out of its general revenues.

Americans have had an extensive and varied experience with private insurance of all kinds, and have been accustomed to make the widest possible use of the principle of insurance in all sorts of business affairs. Health insurance simply means the application of the same principles to matters in which there is an insurable interest on the part of the community or the state. In this type of insurance the application of the principle of mutuality to matters in which the public has an insurable interest means that the direct or incidental benefits to the public are such as to justify the government: (1) in making use of its power to compel persons to insure against specific risks; and (2) in paying, when necessary, part of the cost of such insurance or its expenses of administration and assessing through taxation part of the cost on all of the people.[1]

The experience of Germany and other countries has indicated the preventive values inherent in a health insurance plan.[2] Health insurance will afford a very powerful and

[1] Lindsay, Samuel McCune, "Next Steps in Social Insurance in the United States," *U. S. Bur. Lab. Stat. Mo. Lab. Rev.*, vol. viii, no. 2, Feb., 1919, p. 29.

[2] Fisher, Irving, "Need for Health Insurance," *Amer. Labor Legis. Rev.*, vol. vii, 1917, p. 17.

persuasive stimulus to employers, employees and public men to take fuller and speedier advantage of possible health saving devices and methods of operation.[1]

Applied to the problem of occupational diseases, health insurance should within a comparatively few years make decided gains in the elimination of certain hazardous industrial processes and the prohibition of the use of certain deleterious substances; and at one and the same time, materially improve the health of the workers, reduce mortality, and vitally effect the every-day activities of the people of this country.

[1] In England the discovery of cases of diabetes has been held to be affected by the numbers who come under the National Health Insurance Plan. The same is probably true of other serious ailments. *Statistical Bulletin*, Metropolitan Life Ins. Co., vol. x, no. 11, Nov., 1929, p. 3.

# BIBLIOGRAPHY

## Books

Agricola, Georgius, *De Re Metallica*, translated from the first Latin edition of 1556 by Hoover, Herbert C. and Lou H. (London, 1912). 640 p.

Arlidge, J. T., *The Hygiene, Diseases and Mortality of Occupations* (London, 1892). 568 p.

Ashley, Annie, *The Social Policy of Bismarck* (London, 1912). 95 p.

Aub, J. C., Minot, A. S., Fairhall, L. T. and Resnikoff, P., *Lead Poisoning* (Baltimore, 1926). 265 p.

Bailey, W. F., *A Treatise on the Law of Personal Injuries* (Chicago, 1912). 3 vol.

Beckner, E. R., *A History of Illinois Labor Legislation* (Chicago, 1929). 539 p.

Beddoes, Thomas, *Essay on the Causes, Early Signs and Prevention of Pulmonary Consumption* (London, 1799). 274 p.

Bradbury, Harry B., *Workmen's Compensation Law* (New York, 1917). 1285 p.

Brend, William A., *Health and the State* (London, 1917). 354 p.

Burkitt, Miles C., *Prehistory*, 2nd ed. (Cambridge, 1925). 438 p.

Buxton, L. H. Dudley, *Primitive Labour* (London, 1924). 272 p.

Clark, George L., *The Law of Torts* (Columbia, Mo., 1926). 359 p.

Cohen, Joseph L., *Social Insurance Unified* (London, 1924). 157 p.

——, *Workmen's Compensation in Great Britain* (London, 1923). 232 p.

Commons, J. R. and Andrews, J. B., *Principles of Labor Legislation* (New York, 1920). 559 p.

Cox, Alfred, *Seven Years of National Health Insurance in England* (Chicago, 1921). 43 p.

Dawbarn, C. Y. C., *Employers' Liability to their Servants at Common Law and under the Employers' Liability Act of 1880* (London, 1911), 4th ed. 714 p.

Dublin, Louis I., *Health and Wealth* (New York, 1928). 361 p.

Ellingwood, A. R. and Coombs, W., *The Government and Labor* (New York, 1926). 639 p.

Ferrero, Guglielmo, *The Greatness and Decline of Rome*, translated by Zimmern, Alfred E. (London, 1907). 5 vol.

Freund, Ernst, *The Police Power* (Chicago, 1914). 819 p.

Geikie, James, *Antiquity of Man in Europe* (Edinburgh, 1913). 328 p.

Glotz, Gustave, *The Aegean Civilization* (London, 1925). 422 p.

Goldman, Franz and Grotjahn, Alfred, *Benefits of the German Sickness Insurance System* (Geneva, 1928). 188 p.

Gordon, Alban, *Social Insurance* (London, 1924). 150 p.

Hackett, J. D., *Health Maintenance in Industry* (New York, 1925). 488 p.

Haldane, J. S., Martin, J. S. and Thomas, R. A., *Report on the Health of Cornish Miners* (London, 1904). 107 p.

Hamilton, Alice, *Industrial Poisons in the United States* (New York, 1925). 590 p.

Harper, Samuel A., *The Law of Workmen's Compensation in Illinois* (Chicago, 1920), 2nd ed. 404 p.

Heeren, A. H. L., *Historical Researches in Politics* (London, 1810). 2 vol.

Hill, Leonard, *Caisson Sickness* (London, 1912). 255 p.

Hoffman, F. L., *Facts and Fallacies of Compulsory Health Insurance* (Newark, 1917). 101 p.

Hoffman, F. L., *The Mortality from Cancer Throughout the World* (Newark, 1915). 826 p.

Honnold, A. B., *Treatise on the American and English Workmen's Compensation Acts* (Kansas City, 1917). 2 vol.

Hutchins, B. L. and Harrison, A., *A History of Factory Legislation* (Westminister, 1903). 372 p.

International Labour Office, *Compulsory Sickness Insurance* (Geneva, 1927). 794 p.

——, *Voluntary Sickness Insurance* (Geneva, 1927). 470 p.

Kober, G. M. and Hanson, M., *Diseases of Occupation* (Philadelphia, 1916). 918 p.

Kober, G. M. and Hayhurst, E. R., *Industrial Health* (Philadelphia, 1924). 1184 p.

Lawes, Edward T. H., *The Law of Compensation for Industrial Disease* (London, 1909). 288 p.

Legge, T. M. and Goadby, K. W., *Lead Poisoning and Lead Absorption* (London, 1912). 308 p.

Lloyd George, David, *The People's Insurance* (London, 1911). 161 p.

Macalister, Robert A. S., *A Text Book of European Archaeology* (Cambridge, 1921). 2 vol.

National Bureau of Economic Research, *Income in the United States 1909-1919* (New York, 1922). 2 vol.

National Industrial Conference Board, *Medical Care of Industrial Workers* (New York, 1926). 112 p.

——, *Sickness Insurance or Sickness Prevention?* (New York, 1918), 24 p.

——, *The Work of the International Labour Organization* (New York, 1928). 197 p.

——, *Wages in the United States in 1928* (New York, 1929). 41 p.

——, *Workmen's Compensation Acts in the United States* (New York, 1917). 62 p.

Oliver, Thomas, *Dangerous Trades* (London, 1902). 891 p.

——, *Diseases of Occupation* (New York, 1916). 427 p.

——, *Lead Poisoning* (London, 1914). 294 p.

Pancoast, H. K. and Pendergrass, E. P., *Pneumoconiosis* (New York, 1926). 160 p.

Price, George M., *The Modern Factory* (New York, 1914). 574 p.

Ramazzini, Bernardino, *On the Diseases of Artificers*, translated by James R. (London, 1750). 296 p.

Rubinow, I. M., *Social Insurance* (New York, 1913). 525 p.

——, *Standards of Health Insurance* (New York, 1916). 322 p.

Schrumpf, P., *Tobacco and Physical Efficiency* (New York, 1927). 134 p.

Sherman, P. T., *Criticism of a Tentative Draft of an Act for Health Insurance* (New York, 1917). 93 p.

Smith, Andrew H., *The Effects of High Atmospheric Pressure*, including the *Caisson Disease* (Brooklyn, 1873). 53 p.

Sohm, Rudolph, *Institutes of Roman Law*, translated by Ledlie, J. C. (Oxford, 1907). 606 p.

Spencer, Herbert, *Principles of Sociology* (New York, 1898). 3 vols.

Taylor, Isaac, *The Origin of the Aryans* (London, 1889). 198 p.

Taylor, R. W. C., *The Factory System and the Factory Acts* (London, 1894). 184 p.

Thackrah, Charles T., *The Effects of Arts, Trades and Professions on Health and Longevity* (London, 1832), 2nd ed. 238 p.

Thompson, W. Gilman, *The Occupational Diseases* (New York, 1914). 724 p.

Thompson, W. H., *Workmen's Compensation* (London, 1922). 96 p.

Webb, Mrs. Sidney, *The Case for the Factory Acts* (London, 1902). 233 p.

Weyl, Theodor, *Handbuch der Arbeiterkrankheiten* (Jena, 1908). 809 p.

Woodbury, Robert M., *Social Insurance, an Economic Analysis* (New York, 1917). 171 p.

World Peace Foundation, *Record of the International Labour Organization* (Boston, 1928). 237 p.

Young, Thomas, *A Practical and Historical Treatise on Consumptive Diseases* (London, 1815). 496 p.

### PERIODICALS, PAMPHLETS AND REPORTS

*American Journal of Medical Sciences*, vol. 155 (Philadelphia, Feb., 1918).

*American Journal of Public Health*, vol. xviii, no. 5 (New York, May, 1928); vol. xix, no. 5, May, 1929; vol. xix, no. 6 (June, 1929).

*American Labor Legislation Review*, vol. i (New York, 1911); vol. ii, 1912; vol. iii, 1913; vol. iv, 1914; vol. vi, 1916; vol. vii, 1917; vol. viii, 1918; vol. xv, 1925; vol. xvii, 1927; vol. xviii, 1928.

*Boston Medical and Surgical Journal*, vol. 197, no. 28 (Boston, Jan. 12, 1928).

*British Medical Journal*, vol. ii (London, Sept., 1928).

*California Social Insurance Commission Report* (Sacramento, Jan., 1917, 1919).

*Committee on the Costs of Medical Care, Publication no. 2* (Washington, 1929).

*Connecticut Commission on Public Welfare Report* (Hartford, 1919).

England, *Chief Inspector Factories and Workshops* (London, 1918).

England and Wales, *The Registrar-General's Decennial Supplement*, pt. ii (London, 1921), 1927.

Folks, Homer, *The Distribution of the Cost of Sickness in the United States* (New York, 1928), pamphlet.

*Harpers Magazine*, August, 1929 (New York).

*Illinois Commission on Occupational Diseases Report* (Springfield, Jan., 1911).

*Illinois Health Insurance Commission Report* (Springfield, 1917).

International Labour Office, *Compensation for Occupational Diseases*, Series M, no. 3 (Geneva, 1925).

——, *Industrial and Labour Information*, vol. xxv, no. 5, Jan. 30, 1928.

——, *Occupation and Health Brochure*, nos. 3, 4, 5, 7, 8, 14, 21, 22, 25, 1925; nos. 54, 63, 67, 1926; nos. 77, 78, 1927; nos. 82, 84, 88, 94, 96, 99, 103, 126, 128, 1928.

*The Insurance Almanac* (New York, 1929).

*Journal American Medical Association*, vol. 48, no. 13 (Chicago, March 30, 1907); vol. 60, no. 9, March 1, 1913; vol. 66, no. 1, Jan. 1, 1916.

*Journal of Industrial Hygiene*, vol. i, no. 2 (Baltimore, June, 1919); vol. i, no. 4, Aug., 1919; vol. iii, no. 8, Dec., 1921; vol. iv, no. 4, Aug., 1922; vol. iv, no. 6, Oct., 1922; vol. iv, no. 7, Nov., 1922; vol. iv, no. 8, Dec., 1922; vol. vi, no. 3, July, 1924; vol. vii, no. 1, Jan., 1925; vol. ix, no. 4, April, 1927; vol. ix, no. 8, Aug., 1927; vol. x, no. 4, April, 1928; vol. x, no. 5, May, 1928; vol. x, no. 6, June, 1928; vol. x, no. 9, Nov., 1928; vol. xi, no. 2, Feb., 1929.

*The Lancet*, no. 5487 (London, Oct. 27, 1928); no. 5541, Nov. 9, 1929.

*Massachusetts Special Commission on Social Insurance Report* (Boston, 1917).

*Massachusetts Revised Rules and Regulations Pertaining to the Painting Business* (Boston, Dec., 1925).

Massachusetts Department of Industrial Accidents, *Annual Report for the Year ending June 30, 1927* (Boston, 1927).

Massachusetts Department of Labor and Industry: Division of Industrial Safety, *Report for Year ending Nov. 30, 1928*.

Metropolitan Life Insurance Co., *Statistical Bulletin*, vol. ix, no. 6 (New York, June, 1928); vol. x, no. 11, Nov., 1929; vol. x, no. 12, Dec., 1929.

National Civic Federation, *Report of 17th Annual Meeting* (New York, Feb. 3, 1917).

National Safety Council, *Final Report of the Committee on Spray Coating* (Chicago, 1927).

New Jersey Commission on Old Age, *Insurance and Pensions Report* (Trenton, Nov., 1917).

New Jersey Department of Labor, *The Industrial Bulletin* (Trenton, June, 1929).

New York City Department of Health, *Monograph no. 15.*, Dec., 1915.

——, *Reprint no. 83*, Nov., 1919.

——, *Annual Report*, 1928.

*New York Medical Journal*, vol. 107 (New York, June, 1918).

New York State Factory Investigation Commission *Second Report*, vol. ii (Albany, 1913).

New York State Department of Labor, *Annual Report* (Albany, 1927).

——, *Bulletin no. 83*, July, 1917.

——, *Bulletin no. 90*, Dec., 1918.

——, *Bulletin no. 101*, Dec., 1920.

——, *The Industrial Bulletin*, vol. vi, no. 4, Jan., 1927.

——, *The Industrial Bulletin*, vol. viii, no. 2, Nov., 1928.

——, *The Industrial Bulletin*, vol. viii, no. 3, Dec., 1928.

——, *The Industrial Code Bulletin*, no. 4, effective June 15, 1914.

——, *The Industrial Code Bulletin*, no. 10, effective April 15, 1915.

——-, *The Industrial Code Bulletin*, no. 11, effective April 15, 1915.

——, *The Industrial Code Bulletin*, no. 12, effective May 15, 1915.

——, *The Industrial Code Bulletin*, no. 17, effective July 1, 1918.

——, *The Industrial Code Bulletin*, no. 27, effective Oct. 1, 1924.

——, *The Industrial Code Bulletin*, no. 29, effective Nov. 1, 1926.

——-, *The Industrial Hygiene Bulletin*, vol. viii, no. 10, April, 1927.

——, *The Industrial Hygiene Bulletin*, vol. v, no. 2, Aug., 1928.

——, *The Industrial Hygiene Bulletin*, vol. v, no. 6, Dec., 1928.

——, *The Industrial Hygiene Bulletin*, vol. v, no. 11, May, 1929.

*New York Times*, June 20, 1928; December 21, 1928; January 31, 1930.

New South Wales Director General of Public Health, *Report for 1925*.

*Ohio Health Insurance and Old Age Commission Report* (Columbus, 1919).

Ohio State Board of Health, *Survey of Industrial Health Hazards and Occupational Diseases in Ohio*, Feb., 1915.

*Ontario, Canada Department of Health Annual Report*, 1926.

*Pennsylvania Health Insurance Commission Report* (Harrisburg, Jan., 1919).

*Royal Commission on National Health Insurance Report* (London, 1926).
*Scientific American,* Dec. 25, 1920 (New York).
*Statistic des Deutschen Reichs,* Band 338, Die Krankenversicherung (Berlin, 1927).
*Survey,* Jan. 1, 1930 (New York).
*Union of South Africa Miners' Phthisis Medical Bureau Report* (Pretoria, 1929) ; *Report 1930.*
United States Bureau of Labor Statistics, *Bulletin,* no. 82 (Washington, 1909) ; no. 86, 1910; no. 120, 1913; no. 141, 1914; no. 165, 1915; no. 179, 1915; no. 207, 1917; no. 209, 1917; no. 212, 1917; no. 231, 1918; no. 267, 1920; no. 293, 1922; no. 312, 1923; no. 405, 1926; no. 423, 1926; no. 426, 1927; no. 427, 1927; no. 460, 1928; no. 465, 1928; no. 488, 1929; no. 507, 1930.
United States Bureau of Labor Statistics, *Monthly Labor Review,* vol. ii, no. 6, June, 1916; vol. iii, no. 1, July, 1916; vol. iii, no. 5, Nov., 1916; vol. iv, no. 2, Feb., 1917; vol. iv, no. 4, April, 1917 ; vol. vi, no. 1, Jan., 1918; vol. vi, no. 5, May, 1918; vol. vi, no. 6, June, 1918; vol. viii, no. 1, Jan., 1919; vol. viii, no. 2, Feb., 1919; vol. viii, no. 5, May, 1919; vol. viii, no. 6, June, 1919; vol. ix, no. 1, July, 1919; vol. ix, no. 9, Sept., 1919; vol. ix, no. 6, Dec., 1919; vol. x, no. 2, Feb., 1920; vol. x, no. 4, Apr., 1920; vol. xxi, no. 5, Nov., 1925; vol. xxvi, no. 4, April, 1928; vol. xxvi, no. 6, June, 1928; vol xxvii, no. 1, July, 1928; vol. xxvii, no. 2, Aug., 1928; vol. xxvii, no. 3, Sept., 1928; vol. xxvii, no. 5, Nov., 1928; vol. xxviii, no. 6, June, 1929; vol. xxix, no. 1, July, 1929.
United States Bureau of Mines, *Technical Paper,* no. 105, 1915; no. 260, 1921.
United States Bureau of Mines, *Bulletin no. 132,* 1917.
*United States Census of Manufactures,* 1925.
*United States Census of Occupations,* 1920.
United States Commissioner of Labor, *24th Annual Report,* vol. i, 1909.
*United States Public Health Bulletin,* no. 76, 1916; no. 85, 1917; no. 116, 1921 ; no. 162, 1926; no. 176, 1928.
*United States Public Health Report,* vol. 35, no. 37, Sept. 10, 1920; vol. 37, no. 10, March 10, 1922; vol. 37, no. 50, Dec. 15, 1922; vol. 40, no. 7, Feb. 13, 1925; vol. 43, no. 25, June 22, 1928; vol. 43, no. 29, July 20, 1928; vol. 45, no. 3, Jan. 17, 1930.
United States Secretary of Labor, *Annual Report 1927.*
*Wisconsin Special Commission on Social Insurance Report* (Madison, 1919).
*Wisconsin Labor Department Statistical Bulletin no. 20* (Madison, Sept. 5, 1929).

## Laws

Alabama Code of 1923, sec. 1724.

Alaska Acts of 1917, ch. 4, sec. 2.

California General Laws, Sims' Deerings' Codes 1906, Act no. 1098, sec. 4, as amended 1909, ch. 52.

California Acts of 1913, ch. 81, secs. 1-6.

California Acts of 1913, ch. 186, sec. 1.

California Acts of 1915, ch. 275.

California Acts of 1917, ch. 312.

California Acts of 1917, ch. 586, as amended 1919, ch. 471.

Colorado Compiled Laws 1921, sec. 3482, secs. 4172-4173.

Connecticut Laws 1917, ch. 163.

Connecticut General Statutes 1918, sec. 2350.

Connecticut General Statutes 1918, sec. 5388, as amended 1927, ch. 307, sec. 7.

Connecticut Laws, sec. 2416, as amended 1923, ch. 93.

Delaware Revised Code 1914, ch. 3145, sec. 45, as amended 1923, ch. 202.

Delaware Acts of 1917, ch. 231, sec. 6.

District of Columbia, 45 Stat. 600; 44 Stat. 1424.

Illinois Revised Statutes 1917, ch. 48, secs. 43-46; ch. 48, sec. 99; ch. 48, sec. 154, p. 1469.

Illinois Acts of 1923, p. 352.

Indiana Acts of 1919, ch. 39, sec. 1.

Indiana Acts of 1923, ch. 42, sec. 10.

Kentucky Acts of 1916, ch. 33, sec. 1, as amended 1918, ch. 176; 1922, ch. 50; 1924, ch. 70.

Kentucky Acts of 1919, ch. 162, sec. 2, as amended 1921, ch. 142; 1925, ch. 222.

Louisiana Acts of 1912, Act no. 301, sec. 19.

Maine Revised Statutes 1916, ch. 19, sec. 19.

Maryland Public General Laws, Code of 1911, sec. 5g; added 1912, ch. 165.

Massachusetts Acts of 1916, ch. 33, sec. 1, as amended 1918, ch. 176; 1922, ch. 50; 1924, ch. 70.

Massachusetts Laws of 1917, ch. 130.

Massachusetts General Laws 1921, sec. 142; ch. 149, sec. 54; secs. 117-120; ch. 152.

Michigan Compiled Laws 1915, sec. 5166.

Missouri Revised Statutes 1919, secs. 6817-6818.

Minnesota General Statutes 1913, sec. 3899.

Minnesota Acts of 1919, sec. 20; ch. 84, secs. 6-10.

Minnesota Acts of 1921, ch. 82, pt. 2, sec. 67.

Nevada Acts of 1913, ch. 125, secs. 1-2.

New Hampshire Acts of 1913, ch. 118, sec. 1.

New Mexico Annotated Statutes 1915, sec. 3521.

New Jersey Compiled Statutes 1910, sec. 33, as amended 1912, ch. 5; sec. 35.

New Jersey Acts of 1911, ch. 95; added 1924, ch. 124, sec. 2; ch. 198.

New Jersey Acts of 1914, ch. 121.

New Jersey Acts of 1917, ch. 277.

New York Consolidated Laws 1909, art. ii, secs. 28-29; art. iv, sec. 146, as amended 1921, ch. 642, sec. 147; art. vii, sec. 206; art. xi, sec. 293, para. 2; art. xii, sec. 299; ch. 67, added by 1914, ch. 41; as amended 1920, ch. 538; 1922, ch. 615; 1928, ch. 754; 1929, ch. 298; 1930, ch. 60.

North Dakota Acts of 1919, ch. 162, sec. 2; as amended 1921, ch. 142; 1925, ch. 222.

North Dakota General Laws 1921, ch. 152, sec. 26.

Ohio General Code 1910, sec. 1011; added 1923, p. 314; sec. 6330.

Ohio Laws of 1917, p. 520.

Ohio General Code, secs. 1465-1468a; added 1921, p. 181.

Pennsylvania Statutes 1920, secs. 5424-5436; sec. 13593.

Porto Rico Acts of 1928, art. no. 85, sec. 3.

Rhode Island General Laws 1923, ch. 163, sec. 23.

United States Statutes, Act of April 9, 1912, secs. 6271-6287.

United States Acts of 1915-1916, sec. 40; as amended 1924, ch. 261.

Wisconsin Laws of 1917, ch. 604.

Wisconsin Statutes 1923, sec. 69.49; sec. 102.35; sec. 103.05, para. 3; sec. 110.04.

## TABLE OF CASES CITED

Adams *v.* Acme White Lead and Color Works, 182 Mich. 157, 148 N. W. 485 (1914).

Aetna Life Insurance Co. *v.* Graham, Texas Appeal Comm. Sec. B. 284 S. W. 931 (1926).

Amesbury *v.* Vacuum Oil Co., 9 N. Y. St. Dep. Rep. 399 (1916).

Amsterdam *v.* Hammer Bros., N. Y. S., 210 App. Div. N. Y. 816 (1924).

Anderson *v.* Baxter, Sup. Ct. Penn., Feb., 1926, 132 Atl. 358.

Anderson *v.* Carnegie Steel Co., 255 Penn. 33, 99 Atl. 215 (1916).

Becton *v.* Deas Paving Co., M. and S. La. Digest 154, 3 La. App. 683 (1926).

Bergeron's Case, 243 Mass. 366, 137 N. E. 739 (1923).

Borgnis *v.* Falk Co., 147 Wis. 327, 133 N. W. 209 (1911).

Borgsted *v.* Schultz Bread Co., 167 N. Y. S. 647, 180 App. Div. 229 (1917).

*In re* Bowers, Williams, Colan, 65 Ind. App. 128, 116 N. E. 842 (1917).

Brinton's Ltd. *v.* Turvey, A. C. 230, 7 W. C. C. I. (1905).

Broussard *v.* Union Sulphur Co, 5 La. App. 340, La. Digest 160 (1927).

Burckard *v.* Industrial Comm., 22 Ohio L. R. 420 (Common Pleas), (1924).

Campbell *v.* Industrial Comm. of Ohio, Ohio Ct. of Appeals, 1926, 153 N. E. 276.

Cantor *v.* Elsmere Garage *et al.*, 212 N. Y. S. 327, 214 App. Div. 351 (1925).

Carroll *v.* Industrial Comm. of Colo., 69 Colo. 473, 195 Pac. 1097 (1921).

Chicago Rawhide Mfg. Co. *v.* Industrial Comm., 291 Ill. 616, 126 N. E. 616 (1920).

Chop *v.* Swift and Co., 118 Kan. 35, 233 Pac. 800 (1925).

Cishowski *v.* Clayton Mfg. Co. *et al.*, 105 Conn. 651, 136 Atl. 472 (1927).

Clark *v.* Lehigh Valley Coal Co., 264 Penn. 529, 107 Atl. 858 (1919).

Clinchfield Carbocoal Corp. *v.* Kiser, 139 Va. 387, 124 S. E. 271 (1924).

Columbine Laundry Co. *v.* Industrial Comm., 73 Colo. 397, 215 Pac. 870 (1923).

Cooke *v.* Holland Furnace Co., 200 Mich. 192, 166 N. W. 1013 (1918).

Crooks *v.* Tazewell Coal Co., 263 Ill. 343, 105 N. E. 132 (1914).

*In re* Crowley, 223 Mass. 288, 111 N. E. 786 (1916).

Cunningham *v.* Northwestern Improvement Co., 44 Mont. 180, 119 Pac. 554 (1911).

Curtis-Warner Corp. *v.* Gorman, Sup. Ct. N. Y., Oct., 1925, 130 Atl. 538.

De La Pena *v.* Jackson Stone Co., 103 Conn. 93, 130 Atl. 89 (1925).

Depre *v.* Pacific Coast Forge Co., 145 Wash. 263, 259 Pac. 720 (1927).

Dillingham's Case, Sup. Jud. Ct. Maine, Aug., 1928, 142 Atl. 865.

D'Oliveri *v.* Austin Nichols and Co. *et al.*, 207 N. Y. S. 699, 211 App. Div. 295 (1925).

Dumbluskey *v.* Phila. and Reading Coal and Iron Co., 270 Penn. 22, 112 Atl. 745 (1921).

Dumbrowski *v.* Jennings and Griffin Co., Sup. Ct. Errors Conn., Jan., 1926, 131 Atl. 745.

Dupre *v.* Atlantic Refining Co., 98 Conn. 646, 120 Atl. 288 (1923).

Eldridge *v.* Endicott, Johnson and Co., 189 N. Y. App. Div. 53, 126 N. E. 254 (1920).

Elkhorn Coal Corp. *v.* Kerr, 203 Ky. 804, 263 S. W. 342 (1924).

Ellerman *v.* Industrial Comm., 73 Colo. 20, 213 Pac. 120 (1923).

Farwell *v.* Boston and Worcester R. Co., 4 Metcalf (Mass.) 49 (1842).

Fowler *v.* Risedorph Bottling *et al.*, 161 N. Y. S. 535, 175 N. Y. App. Div. 224 (1916).

Fritz *v.* Elk Tanning Co., 258 Penn. 180, 101 Atl. 958 (1917).

Gay *v.* Hocking Coal Co., 184 Iowa 948, 169 N. W. 360 (1918).

Gibb *v.* New Field By-Products Coal Co., 287 Penn. 300, 135 Atl. 207 (1926).

Gilliland *et al. v.* Ash Grove Lime and Portland Cement Co., 104 Kan. 771, 180 Pac. 793 (1919).

Glennon's Case, 236 Mass. 542, 128 N. E. 942 (1920).

Gordon *v.* Travelers Insurance Co., Texas Civil App., Oct., 1926, 287 S. W. 911.

Guthrie *v.* Detroit Shipbuilding Co., 200 Mich. 355, 167 N. W. 37 (1918).

Hawkins *v.* Bleakly, 243 U. S. 210, 37 Sup. Ct. 255 (1917).

Hayden *v.* Smithville Mfg. Co., 29 Conn. 548 (1861).

Healey's Case, Sup. Jud. Ct. Maine, Sept., 1924, 126 Atl. 21.

Heiers *v.* Hull and Co., 164 N. Y. S. 767, 178 App. Div. 350 (1917).

Heilerman Brewing Co. *v.* Schultz, 161 Wis. 46, 152 N. W. 446 (1915).

Hoag *v.* Kansas Independent Laundry, 113 Kan. 513, 215 Pac. 295 (1923).

Holnagle *v.* Lansing Fuel and Gas Co., 200 Mich. 132, 166 N. W. 843 (1918).

Houston Packing Co. *v.* Mason, Texas Civil App. 1926, 286 S. W. 862.

*In re* Hurle, 217 Mass. 223, 104 N. E. 336 (1914).

Indian Creek Coal and Mining Co. *v.* Calvert, 68 Ind. App. 474, 119 N. E. 519 (1918).

Industrial Comm. *v.* Rice, 26 Ohio App. 497, 160 N. E. 484 (1927).

Industrial Comm. *v.* Russell, 111 Ohio 692, 146 N. E. 305 (1924).

Ives *v.* So. Buffalo R. Co., 201 N. Y. 271, 94 N. E. 431 (1911).

Iwanicki *v.* State Ind. Acc. Comm., 104 Ore. 650, 205 Pac. 990 (1922).

Jakub *v.* Industrial Comm. *et al.*, 288 Ill. 87, 123 N. E. 263 (1919).

Jeffreyes *v.* Charles H. Sager Co., 191 N. Y. S. 354, 198 App. Div. 446 (1921).

Jellico Coal Co. *v.* Adkins, 197 Ky. 684, 247 S. W. 972 (1923).

Jenson *v.* So. Pacific Co., 215 N. Y. 514, 109 N. E. 600 (1915).

*In re* Johnson, 217 Mass. 388, 104 N. E. 735 (1914).

Johnson *v.* London Guarantee and Accident Co., 217 Mass. 388, 104 N. E. 735 (1914).

Johnson *v.* Mary Charlotte Min. Co., 199 Mich. 218, 165 N. W. 650 (1917).

Judice *v.* Degnon Cons. Co., 167 N. Y. S. 1107, 181 N. Y. App. Div. 909 (1917).

Kelly *v.* Watson Coal Co., 272 Penn. 39, 115 Atl. 885 (1922).

Kentucky State Journal Co. *v.* Workmen's Compensation Board, 161 Ky. 562, 172 S. W. 674 (1914).

Knock *v.* Industrial Accident Comm., Sup. Ct. Calif., Feb., 1927, 253 Pac. 712.

Kosick *v.* Manchester Const. Co., 106 Conn. 107, 136 Atl. 870 (1927).

Kovaliski *v.* Collins Co., 102 Conn. 6, 128 Atl. 288 (1925).

Liondale Bleach, Dye and Paint Works *v.* Riker, 85 N. J. 426, 89 Atl. 929 (1914).

Longobardi *v.* Sargent and Co., 100 Conn. 383, 124 Atl. 13 (1924).

Madden's Case, 222 Mass. 487, 111 N. E. 379 (1916).

*In re* Maggelet, 228 Mass. 57, 116 N. E. 972 (1917).

Mailman *v.* Record Co., 118 Maine 172, 106 Atl. 606 (1919).

Malvern Lumber Co. *v.* Sweeney, 116 Ark. 561, 172 S. W. 821 (1914).

Manchline *v.* State Ins. Fund, 279 Penn. 524, 124 Atl. 168 (1924).

Marton *v.* Pittsburg and Lake Erie R. Co., 203 U. S. 284 (1906).

Matis *v.* Schaeffer, 270 Penn. 141, 113 Atl. 64 (1921).

Matthiessen and Hegeler Zinc Co. *v.* Industrial Board, 284 Ill. 378, 120 N. E. 249 (1918).

McGoey *v.* Turin Garage and Supply Co., 186 N. Y. S. 697, 195 App. Div. 436 (1921).

McMurray *v.* Little and Ives Co., 3 N. Y. St. Dep. Rep. 395 (1915).

Meade Fiber Corp. *v.* Starnes, 147 Tenn. 362, 247 S. W. 989 (1923).

Mellquist *v.* Dakota Printing Co., 51 S. D. 359, 213 N. W. 947 (1927).

Mesite *v.* International Silver Co., 104 Conn. 724, 134 Atl. 262 (1926).

Midland Coal Co. *v.* Rucker's Administrator, 211 Ky. 582, 277 S. W. 838 (1925).

Miller *v.* American Steel and Wire Co., 90 Conn. 349, 97 Atl. 345 (1916).

Miller *v.* Director-General of Railroads, 270 Penn. 330, 113 Atl. 373 (1921).

*In re* Mooradjian, 229 Mass. 521, 118 N. E. 951 (1918).

Mountain Timber Co. *v.* Washington, 243 U. S. 219, 37 Sup. Ct. 260 (1917).

Munn *v.* Illinois, 94 U. S. 113 (1878).

Newkirk *v.* Golden Cycle Min. and Red. Co., 79 Colo. 298, 244 Pac. 1019 (1926).

New Marissa Coal Co. *v.* Industrial Comm., 326 Ill. 116, 157 N. E. 32 (1927).

New River Coal Co. *v.* Files, 215 Ala. 64, 109 So. 360 (1926).

New York Central R. Co. *v.* White, 243 U. S. 188, 37 Sup Ct. 247 (1917).

Nicholson *v.* Roundup Coal Mining Co., Sup. Ct. Mont., June, 1927, 257 Pac. 270.

Ocean Accident and Guar. Corp. *v.* Industrial Comm., 66 Utah 600, 245 Pac. 343 (1926).

O'Connell *v.* Adirondack Elec. Power Corp., 185 N. Y. S. 455, 193 App. Div. 582 (1920).

O'Donnell's Case, 237 Mass. 164, 133 N. E. 621 (1921).

Peoria R. Co. *v.* Industrial Bd., 279 Ill. 352, 116 N. E. 651 (1917).

Perkins *v.* Jackson Cushion Spring Co., 206 Mich. 98, 172 N. W. 374 (1919).

Peru Plow and Wheel Co. *v.* Industrial Comm., 311 Ill. 216, 142 N. E. 546 (1924).

Pimental's Case, 235 Mass. 598, 127 N. E. 424 (1920).

Priestly *v.* Fowler, 3 Meeson and Welsby 1, 6 England (1837).

Purchase *v.* Grand Rapids Refrigerator Co., 194 Mich. 103, 160 N. W. 391 (1916).

Puritan Bed Spring Co. *v.* Wolfe, 68 Ind. App. 330, 120 N. E. 417 (1918).

Republic Iron and Steel Co. *v.* Markiowicz, 75 Ind. App. 57, 129 N. E. 710 (1921).

Retmier *v.* Cruse, 67 Ind. App. 192, 119 N. E. 32 (1918).

Richardson *v.* Greenberg, 176 N. Y. S. 651, 188 N. Y. App. Div. 248 (1919).

Rosenthal *v.* National Aniline and Chemical Co., 215 N. Y. S. 621, 216 App. Div. 588 (1926).

Roth *v.* Industrial Comm., 7 Ohio App. 386, 120 N. E. 172 (1918).

Samoskie *v.* Phila. and Reading Coal and Iron Co., 280 Penn. 203, 124 Atl. 471 (1924).

Sayles *v.* Foley, 38 R. I. 484, 96 Atl. 340 (1916).

Slinger *v.* Muskegon Motor Specialties Co., 201 Mich. 473, 167 N. W. 949 (1918).

Smith *v.* International High Speed Tool Co., 98 N. J. 574, 120 Atl. 188 (1923).

Sokol *v.* Stein Fur Dyeing Co. *et al.*, 216 N. Y. S. 167, 216 App. Div. 573 (1926).

Southern Casualty Co. *v.* Flores, Texas Ct. Civil App., March, 1927, 294 S. W. 932.

Standard Cabinet Co. *v.* Landgrove, 76 Ind. App. 593, 132 N. E. 661 (1921).

State *ex. rel.* Davis-Smith Co. *v.* Clausen, 65 Wash. 156, 117 Pac. 1101 (1911).

State *ex. rel.* Rau *v.* District Ct. of Ramsey Co., 138 Minn. 250, 164 N. W. 916 (1917).

State *ex. rel.* Yaple *v.* Creamer, 85 Ohio 349 (1912).

Stombaugh *v.* Peerless Wire Fence Co., 198 Mich. 445, 164 N. W. 537 (1917).

Sugar Co. of Santa Ana *v.* Industrial Acc. Comm., 35 Cal. App. 652, 170 Pac. 630 (1918).

Tarr *v.* Hecla Coal and Coke Co., 265 Penn. 519, 109 Atl. 224 (1921).

Tercillo *v.* Ward Baking Co., 167 N. Y. S. 666, 180 N. Y. App. Div. 302 (1917).

Texas Employers' Insurance Assn. *v.* Jackson, Texas Comm. of App. Sec. B, Nov., 1924, 265 S. W. 1027.

Tintic Milling Co. *v.* Industrial Comm., 60 Utah 14, 206 Pac. 278 (1922).

U. S. Casualty Co. *v.* Matthews, 35 Ga. App. 526, 133 S. E. 875 (1926).

U. S. Fidelity and Guaranty Co. *v.* Industrial Comm., 76 Colo. 263, 230 Pac. 624 (1924).

Van Vleet *v.* Public Service Co. of York, 111 Neb. 51, 195 N. W. 467 (1923).

Victory Sparkler and Specialty Co. *v.* Francks, 147 Md. 368, 128 Atl. 635 (1925).

Vogeley *v.* Detroit Lumber Co., 196 Mich. 516, 162 N. W. 975 (1917).

Voorhees *v.* Smith, Schoonmaker Co., 86 N. J. 500, 92 Atl. 280 (1914).

Wabash R. Co. *v.* Industrial Comm., 286 Ill. 94, 121 N. E. 569 (1918).

Wager *v.* White Star Candy Co., 217 N. Y. S. 173, 217 N. Y. App. Div. 316 (1926).

Wallins Creek Collieries Co. *v.* Williams, 211 Ky. 200, 277 S. W. 234 (1925).

Walsh *v.* River Spinning Co., 41 R. I. 490, 103 Atl. 1025 (1918).

Ward *v.* Beatrice Creamery Co., 104 Okla. 91, 230 Pac. 872 (1924).

Wenrich *v.* Warning, 182 Wis. 379, 196 N. W. 824 (1924).

Western Metal Supply Co. *v.* Pillsbury, 172 Cal. 407, 155 Pac. 491 (1915).

West Side Coal and Mining Co. *v.* Industrial Comm., 321 Ill. 61, 151 N. E. 593 (1926).

Whittle *v.* National Aniline and Chemical Co., 266 Penn. 359, 109 Atl. 847 (1920).

Williams *v.* Missouri Valley Bridge and Iron Co., 212 Mich. 160, 180 N. W. 357 (1920).

Wright *v.* Used Car Exchange, 223 N. Y. S. 245, 221 N. Y. App. Div. 154 (1927).

Young *v.* Western Furniture and Mfg. Co., 101 Neb. 696, 164 N. W. 712 (1917).

Zajkowski *v.* American Steel and Wire Co., 258 Fed. 9 (1918).

# INDEX

267

Bei Fragen zur Produktsicherheit wenden Sie sich bitte an:
If you have any questions regarding product safety,
please contact:

Walter de Gruyter GmbH
Genthiner Straße 13
10785 Berlin
productsafety@degruyterbrill.com